Lecture Notes in Mechanical Engineering

About this Series

Lecture Notes in Mechanical Engineering (LNME) publishes the latest developments in Mechanical Engineering—quickly, informally and with high quality. Original research reported in proceedings and post-proceedings represents the core of LNME. Also considered for publication are monographs, contributed volumes and lecture notes of exceptionally high quality and interest. Volumes published in LNME embrace all aspects, subfields and new challenges of mechanical engineering. Topics in the series include:

- Engineering Design
- Machinery and Machine Elements
- Mechanical Structures and Stress Analysis
- Automotive Engineering
- Engine Technology
- Aerospace Technology and Astronautics
- Nanotechnology and Microengineering
- Control, Robotics, Mechatronics
- MEMS
- Theoretical and Applied Mechanics
- Dynamical Systems, Control
- Fluid Mechanics
- Engineering Thermodynamics, Heat and Mass Transfer
- Manufacturing
- Precision Engineering, Instrumentation, Measurement
- Materials Engineering
- Tribology and Surface Technology

More information about this series at http://www.springer.com/series/11236

Alexander Evgrafov
Editor

Advances in Mechanical Engineering

Selected Contributions from the Conference
"Modern Engineering: Science
and Education", Saint Petersburg, Russia,
June 2014

 Springer

Editor
Alexander Evgrafov
Peter the Great Saint-Petersburg Polytechnic
    University
Saint Petersburg
Russia

ISSN 2195-4356   ISSN 2195-4364   (electronic)
Lecture Notes in Mechanical Engineering
ISBN 978-3-319-29578-7   ISBN 978-3-319-29579-4   (eBook)
DOI 10.1007/978-3-319-29579-4

Library of Congress Control Number: 2016931213

Printed on acid-free paper

This Springer imprint is published by SpringerNature
The registered company is Springer International Publishing AG Switzerland

Preface

The "Modern Engineering: Science and Education" (MESE) conference was initially organized by the Mechanical Engineering Department of Saint Petersburg State Polytechnic University in June 2011 in St. Petersburg (Russia). It was envisioned as a forum to bring together scientists, university professors, graduate students, and mechanical engineers, presenting new science, technology, and engineering ideas and achievements.

The idea of holding such a forum proved to be highly relevant. Moreover, both location and timing of the conference were quite appealing. Late June is a wonderful and romantic season in St. Petersburg—one of the most beautiful cities, located on the Neva river banks, and surrounded by charming greenbelts. The conference attracted many participants, working in various fields of engineering: design, mechanics, materials, etc. The success of the conference inspired the organizers to turn the conference into an annual event.

The third conference, MESE 2014, attracted 140 presentations and covered topics ranging from mechanics of machines, materials engineering, structural strength, and tribological behavior to transport technologies, machinery quality, and innovations, in addition to dynamics of machines, walking mechanisms, and computational methods. All presenters contributed greatly to the success of the conference. However, for the purposes of this book only 19 papers, authored by research groups representing various universities and institutes, were selected for inclusion.

I am particularly grateful to the authors for their contributions and all the participating experts for their valuable advice. Furthermore, I thank the staff and management at the university for their cooperation and support, and especially, all members of the program committee and the organizing committee for their work in preparing and organizing the conference.

Last but not least, I thank Springer for its professional assistance and particularly Mr. Pierpaolo Riva who supported this publication.

Contents

Frameworks not Restorable from Self Stresses 1
Mikhail D. Kovalev

Structural Modifications Synthesis of Bennett Mechanism 9
Munir G. Yarullin and Marat R. Mingazov

Kinematic Research of Bricard Linkage Modifications 17
Munir G. Yarullin and Ilnar A. Galiullin

Drive Selection of Multidirectional Mechanism with Excess Inputs 31
Alexander N. Evgrafov and Gennady N. Petrov

Engineering Calculations of Bolt Connections 39
Alexander A. Sukhanov

**Modern Methods of Contact Forces Between Wheelset
and Rails Determining** 57
Kirill V. Eliseev

**A Novel Design of an Electrical Transmission Line
Inspection Machine** 67
Mohammad Reza Bahrami

One Stable Scheme of Centrifugal Forces Dynamic Balance 75
Vladimir I. Karazin, Denis P. Kozlikin, Alexander A. Sukhanov
and Igor O. Khlebosolov

**New Effective Data Structure for Multidimensional Optimization
Orthogonal Packing Problems** 87
Vladislav A. Chekanin and Alexander V. Chekanin

One-Dimensional Models in Turbine Blades Dynamics 93
Vladimir V. Eliseev, Artem A. Moskalets and Evgenii A. Oborin

**Stationary Oscillation in Two-Mass Machine Aggregate
with Universal-Joint Drive** 105
Vassil Zlatanov

**The Vibrations of Reservoirs and Cylindrical Supports of Hydro
Technical Constructions Partially Submerged into the Liquid** 115
George V. Filippenko

**Mathematical Modelling of Interaction of the Biped Dinamic
Walking Robot with the Ground** 127
Anastasia Borina and Valerii Tereshin

**Programmable Movement Synthesis for the Mobile Robot
with the Orthogonal Walking Drivers** 135
Victor Zhoga, Vladimir Skakunov, Ilya Shamanov and Andrey Gavrilov

**Processing of Data from the Camera of Structured Light for
Algorithms of Image Analysis in Control Systems of Mobile Robots** ... 149
Vladimir Skakunov, Victor Belikov, Victor Zhoga and Ivan Nesmiynov

**Structural and Phase Transformation in Material of Steam
Turbines Blades After High-Speed Mechanical Effect** 159
Margarita A. Skotnikova, Nikolay A. Krylov, Evgeniy K. Ivanov
and Galina V. Tsvetkova

**Stress Corrosion Cracking and Electrochemical Potential
of Titanium Alloys** 169
Vladimir A. Zhukov

Metal Flow Control at Processes of Cold Axial Rotary Forging 175
Leonid B. Aksenov and Sergey N. Kunkin

**Use of the Capabilities of Acoustic-Emission Technique
for Diagnostics of Separate Heat Exchanger Elements** 183
Evgeny J. Nefedyev, Victor P. Gomera and Anatoly D. Smirnov

Author Index 195

Frameworks not Restorable from Self Stresses

Mikhail D. Kovalev

Abstract Some simple hinge trusses in the plain are considered. Generally speaking, they are statically determinate. However, under certain choice of free (not fastened) hinges they allow internal stress. The question of uniqueness of the positions of free hinges, provided the certain self stress and the positions of fastened hinges are known, is investigated.

Keywords Embeddings of pinned graphs · Assur's groups · Self stress · Global rigidity

Introduction

We will focus on the static-geometric properties of ideal flat lever designs. The simplest such designs are flat trusses D_2, D_3 (Fig. 1), and Assur's groups M_3, M_4 [1, 2]. On these schemes the hinges, fixed (pinned) in the plane, are marked by crosses; the free (not fixed) hinges are marked by circles. It is clear how to build inductively trusses D_k and M_k for an arbitrary number k of free joints. The hinges allow all relative rotations in the plane of adjacent levers. There may be combined hinges connecting more than two levers. Considering these designs, we will assume that the positions in the plane of the pinned hinges are not changing. A typical design with the scheme D_k or M_k is a flat truss, that is it does not allow continuous movement of the free joints without changing the lengths of the levers. In addition, it has no internal stresses, i.e. is statically determinate. However, our interest will be focused on special structures allowing internal stresses.

Let p_i be the radius-vector in the plane of the i-th hinge. The equilibrium condition for the forces, applied to a free hinge p_i from the adjacent hinges, looks like

M.D. Kovalev (✉)
Moscow State University, Moscow, Russia
e-mail: kovalev.math@mtu-net.ru

$$\sum_j \omega_{ij}(p_i - p_j) = 0,$$

where the summation is performed on all the hinges adjacent to the i-th hinge, and the scalar $\omega_{ij} = \omega_{ji}$ is called the stress of the lever $p_i\,p_j$. Value ω_{ij} indicate a measure of tension or compression of the lever: if $\omega_{ij} < 0$, the lever is extended, if $\omega_{ij} > 0$, then it is compressed.

In engineering literature the term "stress of the lever" has a different meaning. Namely, stress is a force e_{ij} per unit cross-sectional area of the lever. If s_{ij} is the cross-sectional area of the lever, then the connection of our stress with engineering stress is as follows: $\omega_{ij} = \frac{e_{ij}\, s_{ij}}{|p_i - p_j|}$.

A framework $p = (p_1, p_2, \ldots, p_m)$ with m free hinges [3] is a definite truss or a specified position of the hinge mechanism. If a framework $p = (p_1, p_2, \ldots, p_m)$ in R^2 is specified, the self (or internal) stresses $\omega = \{\omega_{ij}\}$ of p are defined as non-trivial solutions of the homogeneous system of linear equations:

$$\sum_j \omega_{ij}(p_i - p_j) = 0, \quad 1 \le i \le m. \tag{1}$$

If this system has only a trivial solution, then we say that the framework doesn't admit self stresses. Otherwise, the set $\omega = \{\omega_{ij}\}$ of self stresses of p together with the trivial solution of system (1) is a linear space of self stresses of p.

We are interested in the following question: is it possible to restore positions of the free hinges, knowing positions of the pinned hinges and a self stress $\omega = \{\omega_{ij}\}$? If so, then we say that the framework p is restorable from its self stress ω; otherwise we say that p is not restorable from ω. Let a self stress of a framework p be zero on all levers adjacent to a free hinge p_i, then the framework p is not restorable from this self stress. Really, the equilibrium condition of forces does not depend on the position of p_i.

Note also that if the length of a lever is equal to 0 (adjacent hinges coincide), the framework formally allows stress ω^0, non-zero on this lever and equal to zero on all other levers. Such frameworks we call cancellable. Frameworks without coinciding adjacent hinges we call irreducible. A cancellable framework with more than one free hinges is not restorable from its stress ω^0. Indeed, if its free hinge coincides with the adjacent pinned one, then we have another free hinge with all adjacent levers not stressed. If two adjacent free hinges coincide, then their position can be chosen arbitrarily, and the resulting framework will admit the stress ω^0.

Since unstressed levers have no effect while restoring a framework from its self stress, it makes sense to consider internal stresses nonzero on each lever. Frameworks allowing such stress we call completely stressed.

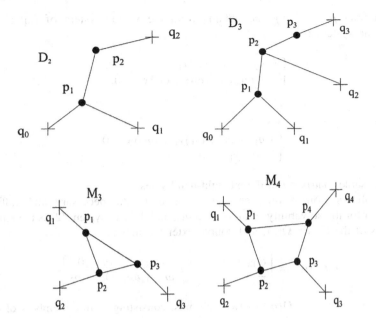

Fig. 1 Hinge schemes D_2, D_3, M_3, M_4

Example 1 Consider a framework with structural scheme D_2 (Fig. 1). Let the positions of the pinned hinges be $q_0 = (0,0)$, $q_1 = (1,0)$, $q_2 = (0,1)$, stresses of adjacent levers be respectively $\omega_0, \omega_1, \omega_2$. In this case the system of Eq. (1) has the form:

$$\begin{cases} \omega_{12}(p_1 - p_2) + \omega_0(p_1 - q_0) + \omega_1(p_1 - q_1) = 0 \\ \omega_{12}(p_2 - p_1) + \omega_2(p_2 - q_2) = 0. \end{cases}$$

If in this system we consider the stresses as known and the radius vectors of free hinges as unknowns, we get the system of equations for the latter:

$$\begin{cases} (\omega_{12} + \omega_0 + \omega_1)p_1 - \omega_{12}p_2 = \omega_0 q_0 + \omega_1 q_1 \\ -\omega_{12}p_1 + (\omega_{12} + \omega_2)p_2 = \omega_2 q_2. \end{cases} \tag{2}$$

The symmetric matrix

$$\begin{bmatrix} \omega_{12} + \omega_0 + \omega_1 & -\omega_{12} \\ -\omega_{12} & \omega_{12} + \omega_2 \end{bmatrix}$$

of the last system is called the matrix of stresses $\Omega(\omega)$.

Assuming $p_i = (x_i, y_i)$ one obtains from the vector system of Eq. (2) two coordinate systems:

$$\begin{cases} (\omega_{12} + \omega_0 + \omega_1)x_1 - \omega_{12}x_2 = \omega_1 \\ -\omega_{12}x_1 + (\omega_{12} + \omega_2)x_2 = 0, \end{cases}$$

and

$$\begin{cases} (\omega_{12} + \omega_0 + \omega_1)y_1 - \omega_{12}y_2 = 0 \\ -\omega_{12}y_1 + (\omega_{12} + \omega_2)y_2 = \omega_2. \end{cases}$$

with the same matrix but different right-hand sides.

As follows from Kronecker-Capelli's theorem the necessary and sufficient condition for the solvability of these systems, and, hence, system (2) is the equality of ranks of the matrix $\Omega(\omega)$ and double-extended matrix

$$\Omega^*(\omega) = \begin{bmatrix} \omega_{12} + \omega_0 + \omega_1 & -\omega_{12} & \omega_1 & 0 \\ -\omega_{12} & \omega_{12} + \omega_2 & 0 & \omega_2 \end{bmatrix}$$

obtained by adding to $\Omega(\omega)$ of two columns, consisting of free members of coordinate systems.

If our framework is completely stressed, the last two columns of matrix $\Omega^*(\omega)$ are independent. Therefore $\det \Omega(\omega) \neq 0$, and system (2) provided its consistency has a unique solution. This means restorability of a completely stressed framework with scheme D_2 from its stress $\omega_0, \omega_1, \omega_2, \omega_{12}$.

Summarizing the reasoning for this example we obtain the following.

Claim 1 *A necessary and sufficient condition for a framework in a plane to be not restorable from its self stress ω is the incompleteness of the rank of the matrix $\Omega(\omega)$ and the equality of this rank to the rank of twice extended matrix $\Omega^*(\omega)$.*

We can prove that for any scheme there exists at least one framework allowing internal stress, and restorable from this stress [4, 5].

Note that if a framework admits self stress, and its fixed hinges all lie on straight line L and free hinges do not all lie on L, it is not restorable from its self stress. This follows from the existence of non-trivial affine transformations of the plane that preserves the points of L stationary, and the following statement. Let framework A (p) be obtained by affine transformation from a framework p.

Claim 2 *Frameworks p and A(p) have the same space of internal stresses.*

The validity of the assertion follows from the fact that both spaces are spaces of solutions of equivalent homogeneous linear systems of the form (1).

From the proposition it also follows that frameworks A(p) and p in this case are both restorable or not restorable from their internal stress ω. Because the non-restorability of p and A(p) from self stress ω is equivalent to the same equality $\det \Omega(\omega) = 0$.

From claim 2 follows:

Claim 3 *If a framework is restorable from its internal stress, and its fixed hinges lie on the same straight line, then all its hinges lie on this line.*

Non Restorable Frameworks for Schemes D_m

For structural schemes D_1 and D_2 (in the case of non-collinear pinned hinges) it is easy to prove that all irreducible completely stressed frameworks are restorable from their self stress. For the scheme D_3 as opposed to D_2 a completely stressed non-restorable framework exists in the case of non-collinear pinned hinges.

Example 2 Consider frameworks with the scheme D_3, and pinned hinges $q_0 = (0,0)$, $q_1 = (1,0)$, $q_2 = (0,1)$, $q_3 = (1,-2)$. The levers of this scheme are $p_1q_0, p_1q_1, p_2q_2, p_3q_3, p_1p_2, p_2p_3$. It turns out that a framework with free hinges $p_1^0 = (1,-1)$, $p_2^0 = (-1,2)$, $p_3^0 = (0,0)$ is not restorable from its self stress $(2,1,1,1,-1,1)$ (the coordinates of the vector correspond to the order listed in above levers). Here, the twice extended matrix looks like this:

$$\begin{bmatrix} 2 & 1 & 0 & 1 & 0 \\ 1 & 1 & -1 & 0 & 1 \\ 0 & -1 & 2 & 1 & -2 \end{bmatrix},$$

and its rank is equal to two as well as for the stress matrix. Each of the frameworks $p_1 = p_1^0 - X$, $p_2 = p_2^0 + 2X$, $p_3 = p_3^0 + X$, where X is a arbitrary plane vector, also allows internal stress $(2,1,1,1,-1,1)$.

Note also that framework $p_1^0 = (1,0), p_2^0 = (-1,0), p_3^0 = (0,-1)$ from this linear manifold is cancellable and has a two-dimensional space of internal stresses $(2u, v, u, u, -u, u)$, $u, v \in R$. This framework is restorable from all of it's stress with $u \neq 0$ and $u \neq v$, for example, from stress $(2,2,1,1,-1,1)$.

Thus, the cancellable framework with two-dimensional space of self stresses, restorable from some of its stresses and not restorable by other self stresses is pointed out. It is also clear that if there is such a property, then the set of non-restoring stress is a closed subset of incomplete dimension of the set of all self stresses of the framework. Note also that for the scheme D_3 with fixed hinges $q_0 = (0,0)$, $q_1 = (1,0)$, $q_2 = (0,1)$, $q_3 = (A,B)$ for a dense open set of parameters, A, B there exist irreducible completely stressed frameworks not restorable from their one-dimensional space of internal stresses.

For this scheme there are possible pinnings, for which all irreducible completely stressed frameworks are restorable from their stress.

Theorem 1 *For scheme D_3 all irreducible frameworks, admitting a self stress non-zero on all levers, are restorable from the stress at the following pinnings:*

Fig. 2 Scheme T_3

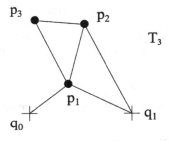

1. $q_2 = q_3$ and $q_0 = q_1$.
2. *Exactly three pinned hinges coincide.*
3. *q_0, q_1, q_2 are collinear, but q_0, q_1, q_2, q_3 are not collinear.*
4. *q_0, q_1, q_2 are not collinear, and q_0, q_2, q_3 are collinear* (Fig. 2).

Consider a scheme T_m obtained by a sequence of additions of groups, consisting of a hinge and two adjacent levers to the adjacent hinges, beginning with the simplest scheme, consisting of one free hinge p_1 and two adjacent pinned hinges q_0, q_1. For example, a group of the levers $p_2 p_1, p_2 q_1$, and then a group $p_3 p_2, p_3 p_1$, and so on. For the scheme T_m a completely stressed framework is obtained only when all its hinges are on the line $q_0 q_1$. And such an irreducible framework is restorable from each of its internal non-zero on all levers stress, as follows from theorems of paper [4]. In this case the set of completely stressed frameworks has dimension equal to m, a half of the dimension of the set of all frameworks.

Theorem 2 *For a scheme $D_m, m \geq 3$ for an arbitrary choice of positions of pinned hinges, excluding one of hinges q_k, $k = 2, 3$, there exists an irreducible completely stressed framework not restorable from its internal stress.*

Non Restorable Frameworks for Schemes M_m

In [4] it was found that if pinned hinges of scheme M_3 (Fig. 1) do not lie on the same straight line, then all completely stressed irreducible frameworks with this scheme are restorable. However, for the scheme M_4 there are completely stressed irreducible frameworks not restorable from self stress, even with pinned hinges not collinear.

Theorem 3 *For the scheme M_4 for any pinning, excluding the case when the hinges do not lie on a straight line and two of them coincide, there is a framework not restorable from its self stress non-zero on all the levers.*

Theorem 3 implies that for scheme M_4 all irreducible frameworks admitting self stress non-zero on all the levers are restorable from this self stress only when the pinned hinges do not lie on a straight line and two of them coincide.

In the case of a larger number of hinges the schemes M_m are also Assur's groups [6, 7]. The situation for them is similar to that for the schemes D_m.

Theorem 4 *For any hinged scheme M_m, $m \geq 4$ for an arbitrary choice of positions of pinned hinges, excluding two of them, there exists an irreducible completely stressed framework not restorable from its internal stress.*

If a flat framework p^0 is not restorable from its internal stress ω, then a linear manifold $L(\omega)$ of frameworks adopting this stress exists. The dimension of $L(\omega)$ is equal to 2k, where k is corank of matrix $\Omega(\omega)$. As shown in the following example, there may be a manifold $L(\omega)$ for ω nonzero on all levers, with any framework having constant position of a free hinge.

Example 3 Consider framework p^0 with scheme M_4 with pinned hinges $q_1 = (0,0), q_2 = (1,0), q_3 = \left(\frac{1}{2}, \frac{1}{2}\right), q_4 = (0,1)$ and free hinges $p_1^0 = \left(\frac{1}{4}, \frac{1}{2}\right)$, $p_2^0 = p_4^0 = (0,0)$, $p_3^0 = \left(\frac{1}{2}, \frac{1}{2}\right)$. The framework p^0 is not restorable from its self stress $\omega_1 = 1$, $\omega_2 = \frac{1}{4}$, $\omega_3 = 2$, $\omega_4 = \frac{1}{2}$, $\omega_{12} = 1$, $\omega_{23} = -1$, $\omega_{34} = 1$, $\omega_{14} = -2$. (Here we have designated ω_i the stress of a lever $p_i q_i$.) The positions of the free hinges of frameworks admitting this internal stress are as follows: $p_1 = p_1^0, p_2 = 2X, p_3 = p_3^0 - \frac{1}{2}X, p_4 = X$, where X—is an arbitrary vector in a plane.

The following theorem is valid.

Theorem 5 *For the scheme M_4 for the following choice of pinned hinges: q_1, q_2, q_4 are not collinear, and q_3 lies on the line q_2, q_4 in the variety $L(\omega)$ of frameworks the free hinge p_1 has an unchangeable position.*

References

1. Assur LV (1952) Investigation of flat lever mechanisms with lower kinematic pairs from the point of view of their structure and classification. Izd. AN SSSR, 589 p. (In Russian)
2. Kovalev MD (2006) On structural Assur's groups. Teoriya Mehanizmov i Mashin. № 1 (7) 4, 18–26. (In Russian)
3. Kovalev MD (1994) Geometric theory of hinged devices//Izvestiay RAN Seriya matematicheskaya 58(1):45–70
4. Kovalev MD (1997) On the reconstructibility of frameworks from self-stresses//Izvestiay RAN Seriya matematicheskaya 61(4):37–66
5. Kovalev MD (2001) Questions of the geometry of hinged devices and schemes//Vestnik MGTU. Seriya Mashinostroenie. № 4, pp 33–51. (In Russian)
6. Servatius B, Shai O, Whiteley W (2010) Combinatorial characterization of the assur graphs from engineering. Eur J Comb 31(4):1091–1104
7. Servatius B, Shai O, Whiteley W (2010) Geometric properties of assur graphs. Eur J Comb 31 (4):1105–1120

Structural Modifications Synthesis of Bennett Mechanism

Munir G. Yarullin and Marat R. Mingazov

Abstract The article presents 16 models of Bennett mechanism modifications depending on location of twisted angles in different quadrants. Two types of mechanism are identified—"parallelogram" and "isogram" of Bennett's mechanism. Based on the results of system analysis research, it was established that if Bennett mechanism twisted angles are located in adjacent quadrants, then such a mechanism would be a parallelogram Bennett. If Bennett mechanism twisted angles are located in one quadrant or in the opposite of quadrants, that such a mechanism would be an "isogram" of Bennett.

Keywords Spatial mechanisms · Structure · Mobility of mechanism · Joint · Revolute pair · Skew axes · Crank · Rod

Introduction

Bennett mechanism is a spatial four-bar mechanism, firstly described by English mathematician Bennett in 1903 [1]. Many authors have conducted researches of Bennett mechanism [2–9]. Relationship between rotations of input link to output link is described in papers [5, 10]. Although the spatial four-bar mechanism is one of the simplest mechanisms capable of transmitting to rotation movement from one plane to another, but in practice it is not yet widely used.

M.G. Yarullin (✉) · M.R. Mingazov
Kazan National Research Technical University, Tatarstan, Russia
e-mail: Yarullinmg@yahoo.com

M.R. Mingazov
e-mail: maratmingazovr@gmail.com

© Springer International Publishing Switzerland 2016
A. Evgrafov (ed.), *Advances in Mechanical Engineering*,
Lecture Notes in Mechanical Engineering, DOI 10.1007/978-3-319-29579-4_2

9

Fig. 1 Structure scheme of spatial four-bar mechanism

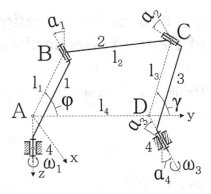

The Bennett mechanism is the only movable spatial 4R mechanism with non-parallel and non-intersecting joint axes (Fig. 1). According to Grübler-Kutzbach's mobility criterion, mechanism has negative mobility:

$$W = 6 \cdot (m - 1) - 5 \cdot p = 6 \cdot (4 - 1) - 5 \cdot 4 = -2, \tag{1}$$

where m is the number of links, p is the number of revolute pairs.

However, in practice, the mechanism has a single mobility, but to achieve this result, it is necessary to provide accurately agreed size:

1. opposite links must be twisted the same amount ($\alpha_1 = \alpha_3, \alpha_2 = \alpha_4$),
2. opposite links must have equal lengths ($l_1 = l_3, l_2 = l_4$),
3. the link-twists and link-lengths must be related as:

$$\frac{l_1}{\sin \alpha_1} = \pm \frac{l_2}{\sin \alpha_2} \tag{2}$$

Bennett Mechanism Modifications

The sign $\pm$ in Eq. (2) indicates that there are 4 different types of this equation:

$$+ \frac{l_1}{\sin \alpha_1} = + \frac{l_2}{\sin \alpha_2}, \tag{3}$$

$$+ \frac{l_1}{\sin \alpha_1} = - \frac{l_2}{\sin \alpha_2}, \tag{4}$$

$$- \frac{l_1}{\sin \alpha_1} = + \frac{l_2}{\sin \alpha_2}, \tag{5}$$

$$-\frac{l_1}{\sin\alpha_1} = -\frac{l_2}{\sin\alpha_2}. \tag{6}$$

The parameters l_1, l_2 are length of links, both $l_1 \geq 0$ and $l_2 \geq 0$ because the length of the links cannot be negative. The parameters α_1, α_2 are twisted angles and $(-1 \leq \sin\alpha_1 \leq 1$ and $-1 \leq \sin\alpha_2 \leq 1)$.

According to the trigonometric formulas, we have:

$$\sin(\pi - \alpha) = \sin(\alpha), \tag{7}$$

$$\sin(\pi + \alpha) = -\sin(\alpha), \tag{8}$$

$$\sin(2\pi + \alpha) = \sin(\alpha), \tag{9}$$

$$\sin(2\pi - \alpha) = -\sin(\alpha). \tag{10}$$

So, Eqs. (7)–(10) indicate that the mechanism may have different modifications depending on location of parameters α_1, α_2 in different quadrants. In our previous paper [8] we provide a brief analysis of spatial 4R mechanism modifications. This paper is a continuation of it and devoted to definition of 16 mechanism modifications depending on location of twisted angles in different quadrants. For each modification we define the type of mechanism (parallelogram or isogram) and the relationship between rotations of input crank to output crank.

Thus, according to Eqs. (11)–(14) we can represent Eq. (2) as follow:

$$\frac{l_1}{l_2} = \frac{\sin\alpha_1}{\sin\alpha_2}, \tag{11}$$

$$\frac{l_1}{l_2} = \frac{\sin\alpha_1}{\sin(\pi - \alpha_2)} = \frac{\sin\alpha_1}{\sin\alpha_2'}, \tag{12}$$

$$\frac{l_1}{l_2} = \frac{\sin\alpha_1}{\sin(\pi + \alpha_2)} = \frac{\sin\alpha_1}{\sin\alpha_2''}, \tag{13}$$

$$\frac{l_1}{l_2} = \frac{\sin\alpha_1}{\sin(2\pi - \alpha_2)} = \frac{\sin\alpha_1}{\sin\alpha_2'''}, \tag{14}$$

$$\frac{l_1}{l_2} = \frac{\sin(\pi - \alpha_1)}{\sin\alpha_2} = \frac{\sin\alpha_2'}{\sin\alpha_2}, \tag{15}$$

$$\frac{l_1}{l_2} = \frac{\sin(\pi - \alpha_1)}{\sin(\pi - \alpha_2)} = \frac{\sin\alpha_1'}{\sin\alpha_2'}, \tag{16}$$

$$\frac{l_1}{l_2} = \frac{\sin(\pi - \alpha_1)}{\sin(\pi + \alpha_2)} = \frac{\sin\alpha_1'}{\sin\alpha_2''}, \tag{17}$$

$$\frac{l_1}{l_2} = \frac{\sin(\pi - \alpha_1)}{\sin(2\pi - \alpha_2)} = \frac{\sin \alpha_1'}{\sin \alpha_2'''}, \tag{18}$$

$$\frac{l_1}{l_2} = \frac{\sin(\pi + \alpha_1)}{\sin \alpha_2} = \frac{\sin \alpha_1''}{\sin \alpha_2}, \tag{19}$$

$$\frac{l_1}{l_2} = \frac{\sin(\pi + \alpha_1)}{\sin(\pi - \alpha_2)} = \frac{\sin \alpha_1''}{\sin \alpha_2'}, \tag{20}$$

$$\frac{l_1}{l_2} = \frac{\sin(\pi + \alpha_1)}{\sin(\pi + \alpha_2)} = \frac{\sin \alpha_1''}{\sin \alpha_2''}, \tag{21}$$

$$\frac{l_1}{l_2} = \frac{\sin(\pi + \alpha_1)}{\sin(2\pi - \alpha_2)} = \frac{\sin \alpha_1''}{\sin \alpha_2'''}, \tag{22}$$

$$\frac{l_1}{l_2} = \frac{\sin(2\pi - \alpha_1)}{\sin \alpha_2} = \frac{\sin \alpha_1'''}{\sin \alpha_2}, \tag{23}$$

$$\frac{l_1}{l_2} = \frac{\sin(2\pi - \alpha_1)}{\sin(\pi - \alpha_2)} = \frac{\sin \alpha_1'''}{\sin \alpha_2'}, \tag{24}$$

$$\frac{l_1}{l_2} = \frac{\sin(2\pi - \alpha_1)}{\sin(\pi + \alpha_2)} = \frac{\sin \alpha_1'''}{\sin \alpha_2''}, \tag{25}$$

$$\frac{l_1}{l_2} = \frac{\sin(2\pi - \alpha_1)}{\sin(2\pi - \alpha_2)} = \frac{\sin \alpha_1'''}{\sin \alpha_2'''}. \tag{26}$$

The equations presented above are variations of Eq. (2) depending on the location of twisted angles α_1, α_2 in one of the four quadrants (the order of the quadrants is shown in Fig. 2a). Let's assume that $0° \le \alpha_1 \le 90°$ and $0° \le \alpha_2 \le 90°$. So, the Eq. (7) corresponds to the location of twisted angles shown in Fig. 2b (both angles α_1 and α_2 are located in the first quadrant). The Eq. (8) corresponds to the location of twisted angles shown in Fig. 2c (angle α_1 located on the first quadrant and angle α_2 on the second quadrant).

All possible variants of location of twisted angles according to Eqs. (7)–(22) are shown in Fig. 3. This picture presents 16 items. Each item displays locations of twisted angles in four quadrants.

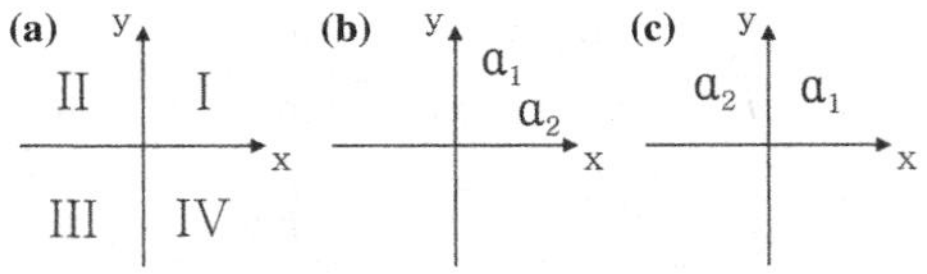

Fig. 2 Location of the twisted angles

Fig. 3 Variants of location of the twisted angles

$$
\begin{array}{c|c|c|c|c}
\begin{matrix}\alpha_1\\\alpha_2\end{matrix} & \alpha'_2 \;\; \alpha_1 & \alpha_1 & \alpha_1 & \alpha_1 \\
 & & \alpha''_2 & & \alpha'''_2 \\
\hline
\alpha'_1 \;\; \alpha_2 & \begin{matrix}\alpha'_1\\\alpha'_2\end{matrix} & \alpha'_1 & \alpha'_1 & \\
 & & \alpha''_2 & & \alpha'''_2 \\
\hline
\alpha_2 \;\; \alpha'_2 & & & & \\
\alpha''_1 & \alpha''_1 & \begin{matrix}\alpha''_1\\\alpha''_2\end{matrix} & \alpha''_1 & \alpha'''_2 \\
\hline
\alpha_2 \;\; \alpha'_2 & & & & \\
\alpha'''_1 & \alpha'''_1 & \alpha''_2 \;\; \alpha'''_1 & \alpha'''_1 & \begin{matrix}\alpha'''_1\\\alpha'''_2\end{matrix}
\end{array}
$$

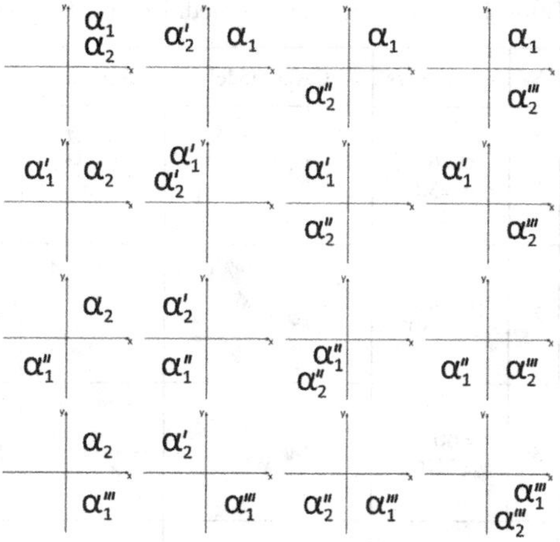

Fig. 4 Spatial 4R mechanism with $l_1 > l_2$

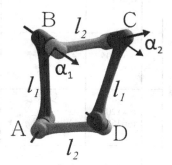

Let the spatial 4R linkage be constructed with structure parameters $l_1, \alpha_1, l_2, \alpha_2$, where $0° \le \alpha_1 \le 90°, 0° \le \alpha_2 \le 90°, l_1 > l_2$ (Fig. 4).

So, from trigonometric transformations we can get 16 modifications of 4R linkage depending on location of the twisted angles in different quadrants (Table 1).

For the study of kinematics modifications of Bennett's mechanisms as presented above, they represent a kinematic scheme of the mechanism in a closed vector loop $\overline{ABCDA}$ as follow:

$$\overline{AB} + \overline{BC} + \overline{CD} = \overline{AD}.$$

It is projected on the axis x, y and z as follows:

$$
\begin{cases}
AB \cdot K_X^{AB} + BC \cdot K_X^{BC} + CD \cdot K_X^{CD} = AD \cdot K_X^{AD} \\
AB \cdot K_Y^{AB} + BC \cdot K_Y^{BC} - CD \cdot K_Y^{CD} = AD \cdot K_Y^{AD} \\
AB \cdot K_Z^{AB} + BC \cdot K_Z^{BC} - CD \cdot K_Z^{CD} = AD \cdot K_Z^{AD}
\end{cases}.
$$

Table 1 Bennett mechanism modifications l_1(const) > l_2(const)

№	parameters	CAD model	Scheme	№	parameters	CAD model	Scheme
1	$\alpha_1=60°$ $\alpha_2=30°$		α_1 α_2	9	$\alpha_1=240°$ $\alpha_2=30°$		α_2 α_1''
2	$\alpha_1=60°$ $\alpha_2=150°$		α_2' α_1	10	$\alpha_1=240°$ $\alpha_2=150°$		α_2' α_1''
3	$\alpha_1=60°$ $\alpha_2=210°$		α_1 α_2''	11	$\alpha_1=240°$ $\alpha_2=210°$		α_1'' α_2''
4	$\alpha_1=60°$ $\alpha_2=330°$		α_1 α_2'''	12	$\alpha_1=240°$ $\alpha_2=330°$		α_1'' α_2'''
5	$\alpha_1=120°$ $\alpha_2=30°$		α_1' α_2	13	$\alpha_1=300°$ $\alpha_2=30°$		α_2 α_1'''
6	$\alpha_1=120°$ $\alpha_2=150°$		α_1' α_2'	14	$\alpha_1=300°$ $\alpha_2=150°$		α_2' α_1'''
7	$\alpha_1=120°$ $\alpha_2=210°$		α_1' α_2''	15	$\alpha_1=300°$ $\alpha_2=210°$		α_2'' α_1'''
8	$\alpha_1=120°$ $\alpha_2=330°$		α_1' α_2'''	16	$\alpha_1=300°$ $\alpha_2=330°$		α_1''' α_2'''

Given that $AB = CD = l_1$, $BC = AD = l_2$ and corresponding direction cosines, we get:

$$\begin{cases} -l_1 \cdot \sin\varphi + l_2 \cdot \sin\gamma \cdot \cos\alpha_1 \cdot \cos\varphi - l_2 \cdot \cos\gamma \cdot \sin\varphi = l_1 \cdot \sin\gamma \cdot \cos\alpha_2 \\ l_1 \cdot \cos\varphi + l_2 \cdot \sin\gamma \cdot \cos\alpha_1 \cdot \sin\varphi + l_2 \cdot \cos\gamma \cdot \cos\varphi = l_2 + l_1 \cdot \cos\gamma \\ l_2 \cdot \sin\alpha_1 = l_1 \cdot \sin\alpha_2 \end{cases}.$$

We have solved the system of equations for $\cos\gamma$ and $\sin\gamma$ and found:

$$\cos\gamma = \frac{K_1 \cdot \cos\varphi - K_2}{K_1 - \cos\varphi \cdot K_2}, \tag{27}$$

$$\sin\gamma = \frac{\sin\varphi \cdot K_3}{K_1 - \cos\varphi \cdot K_2}, \tag{28}$$

where:

$$K_1 = l_2^2 \cdot \cos\alpha_1 + l_1^2 \cdot \cos\alpha_2,$$

$$K_2 = l_1 \cdot l_2 \cdot (\cos\alpha_1 + \cos\alpha_2),$$

$$K_3 = l_2^2 - l_1^2.$$

Based on the results of the CAD models analysis, it was established that if Bennett mechanism twisted angles are located in adjacent quadrants, that such a mechanism would be a parallelogram Bennett (Fig. 5a). If the twisted angles are located in same quadrants or in the opposite quadrants, that such a mechanism would be an "isogram" of Bennett (Fig. 5b).

Based on results of Eqs. (23), (24) calculations establish that, if the mechanism has a "parallelogram" structure, then input and output cranks rotate in the same direction (Fig. 6a). Otherwise, if a mechanism has an "isogram" structure, then input and output cranks rotate in opposite directions with respect to the clockwise (Fig. 6b).

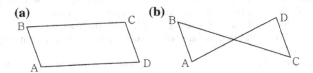

Fig. 5 Structure. **a** Parallelogram, **b** isogram

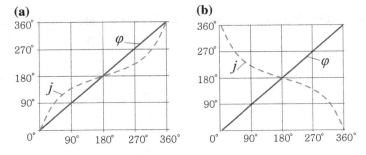

Fig. 6 Input and output link rotation. **a** Same direction, **b** opposite direction

Conclusion

This work identified the 16 modifications of spatial 4R mechanisms depending on location of twisted angles in different quadrants. For each modification the type of mechanism (parallelogram or isogram) and the relationship between rotation of input crank to output crank was defined. It was established, that:

1. If Bennett mechanism twisted angles are located in adjacent quadrants, then such a mechanism would be a parallelogram Bennett.
2. If Bennett mechanism twisted angles are located in one quadrant or in the opposite of quadrants, then such a mechanism would be an "isogram" of Bennett.

Acknowledgements This work was supported by the Russian Foundation for Basic Research (project No 13-08-97090\14).

References

1. Bennett GT (1903) A new mechanism. Engineering 76:777–778
2. 100 years to Bennett mechanism (2004) Proceedings of the international conference on the theory of machines and mechanisms. RIC "Shcola", Kazan, 292 p. (In Russian)
3. Dvornikov LT (2009) Non-traditional arguments about the existence of Bennett mechanism. Theor Mach Mech 1(13):5–10. (In Russian)
4. Mingazov MR, Yarullin MG (2012) Analysis of studies of spatial mechanisms with revolute pairs. Analytical Mechanics, Stability and Control. In: Proceedings of the international X Chetaev conference. Isd-vo Kazan. Gos. Tech. Un-ta, Kazan, pp 358–366. (In Russian)
5. Mingazov MR (2013) Input crank kinematics of spatial 4R linkage. International youth scientific conference XXI Tupolev reading, pp 214–218. (In Russian)
6. Mudrov PG (1976) Spatial mechanisms with revolute joints. Isd-vo Kazanskogo gos. un-ta, Kazan, 264 p. (In Russian)
7. Mudrov AG (2003) Spatial mechanisms with special structure. Shkola, Kazan, 300 p. (In Russian)
8. Yarullin MG, Mingazov MR (2012) A brief analysis of Bennett mechanism modifications Problems of modern machines. Isd-vo VSGUTU, Ulan-Ude 1, p 177–181 (In Russian)
9. Baker JE (1988) The bennett linkage and its associated quadric surfaces. Mech Mach Theor 23:147–156
10. Bennett GT (1913) The skew isograms mechanism. Proc. Lond. Math. Soc. 2 s 13:151–173

Kinematic Research of Bricard Linkage Modifications

Munir G. Yarullin and Ilnar A. Galiullin

Abstract This article considers the problem of Bricard linkage and its modifications kinematic research. This linkage is interesting for its paradoxical mobility, which does not match with the value obtained by using common methods. This feature makes it possible to find practical usages of Bricard linkage in different areas. The paper shows application of the matrix transformations method to determine the kinematic parameters of Bricard linkage and its modifications. The article describes principles of choosing the coordinate systems associated with the joints to determine linkage's kinematic parameters. Using matrix transformation method allowed us to obtain the system of equations describing the relationship between the rotation angles of the mechanism links. As a result, this research gives formulas of angular velocity and acceleration for linkage's joints.

Keywords Bricard linkage · Kinematics · Overconstrained linkages · Transformation matrix · Bricard linkage modifications · Angular velocity · Angular acceleration

Introduction

The "classical Bricard linkage", proposed in 1927 [1], is presented in Fig. 1. This mechanism contains six identical links (AB, BC, CD, DE, EF, FA), connected by revolute joints (A, B, C, D, E, F). The AB link of this mechanism is accepted as its frame and the BC link as its driving link. Theoretical calculation of degree of freedom for spatial mechanisms is described in different researches [2–6]. However, the Bricard linkage has one degree of freedom, which does not match with the zero value, obtained by using common methods [7]. This paradoxal mobility makes it

M.G. Yarullin (✉) · I.A. Galiullin
Kazan National Research Technical University, Kazan, Russia
e-mail: Yarullinmg@yahoo.com

I.A. Galiullin
e-mail: TGaliullin@gmail.com

© Springer International Publishing Switzerland 2016
A. Evgrafov (ed.), *Advances in Mechanical Engineering*,
Lecture Notes in Mechanical Engineering, DOI 10.1007/978-3-319-29579-4_3

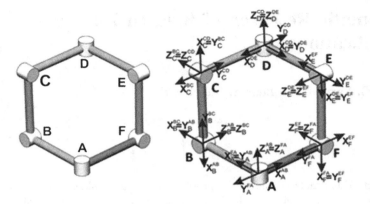

Fig. 1 The Bricard linkage and coordinate systems

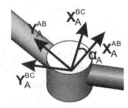

Fig. 2 Angle between X axes of joint A coordinate systems

possible to find practical usages of Bricard linkage in different areas [8–11]. Thus, the development of a new method of Bricard linkage kinematic analysis is the actual task.

To analyze kinematic parameters of this mechanism's links, as well as its characteristic points, it is necessary to define coordinate systems for each joint of the mechanism. To explain the principle of the coordinate systems, we look at Fig. 2 and consider the link AB.

The AB link of Bricard linkage contains two joints (joint A and joint B in Fig. 2). The first coordinate system is placed at the center of joint A. The axis Z_A is collinear with the joint's rotation axis. The positive direction of axis Z of the first link must be selected arbitrarily. The Y_A axis is placed along the link in the direction towards joint B. Axis X_A is placed so that vectors X_A, Y_A and Z_A form a right-handed coordinate system.

Similarly, the second coordinate system is placed at the center of joint B. The axis Z_B is collinear with the joint B rotation axis. The positive direction of axis Z_B must be selected so that it rotates clockwise when moving from joint A to joint B. The Y_B axis is placed along the link in the direction opposite to the A joint. Finally, the axis X_B is placed so that vectors X_B, Y_B and Z_B form a right-handed coordinate system.

Next, consider connection of links AB and BC. Joint B contains two coordinate systems:

X_B^{AB}, Y_B^{AB}, Z_B^{AB}—coordinate system of joint B, as part of AB link,
X_B^{BC}, Y_B^{BC}, Z_B^{BC}—coordinate system of joint B, as part of BC link.

Axes Z_B^{BC} and Z_B^{AB} of joint B are collinear. The Y_B^{AB} axis is placed along the link with its direction towards the A joint. The Y_B^{BC} axis is placed along the link by the opposite direction to the C joint. Axes X_B^{AB} and X_B^{BC} are placed so that other vectors form a right-handed coordinate system.

Therefore, each joint of Bricard linkage contains two coordinate systems. The Bricard linkages have 12 coordinate systems, as shown in Fig. 2.

The angle between X axes of joint A coordinate systems is marked as α_A (see Fig. 2).

Similarly:
α_B—angle between X axes of joint B coordinate systems, α_C—angle between X axes of joint C coordinate systems, α_D—angle between X axes of joint D coordinate systems, α_E—angle between X axes of joint E coordinate systems, α_F—angle between X axes of joint F coordinate systems.

The BC link of this mechanism is accepted as the driving link. Therefore, α_B is the driving link rotation angle. The α_B value is the input parameter of this system. In this research, we determine dependency between angles α_C, α_D, α_E, α_F, α_A and driving link rotation angle α_B.

Transformation Matrix

Computer graphics algorithms use matrices to determine spatial positions of objects. The transformation matrix allows us to define an object's position change.

The position of an object is defined relative to some single world coordinate system (CS_W), which can be arbitrarily selected by the developer. Each object has a local coordinate system (CS_L), hard-connected with another object. So, the spatial position of any object can be defined by the position of CS_L relative to the position of CS_W.

The position of one coordinate system relative to another can be described by the following parameters: position of zero point of the local coordinate system relative to the world coordinate system (offset); direction of vector X of the local coordinate system; direction of vector Y of the local coordinate system and direction of vector Z of the local coordinate system.

Fig. 3 Coordinate systems of some object

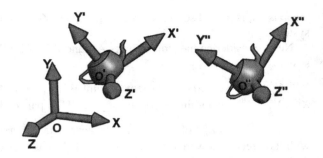

Figure 3 shows CS_W (XYZ) with center at point O and some object with hard-connected CS_L (X'Y'Z') with center at point O'. The position of this object can be completely defined by the position of CS_L (X'Y'Z') relative to CS_W (XYZ).

The position of CS_L (X'Y'Z') relative to CS_W (XYZ) can be completely defined by the defining position of zero point O' relative to CS_W (offset) and projections of vectors X'Y'Z' to the CS_W.

Thus, position of some object in the world coordinate system can be completely defined by a transformation matrix from the CS_W (XYZ) to the CS_L (X'Y'Z'):

$$M_{OO'} = \begin{pmatrix} V_{OX}^{O'X'} & V_{OY}^{O'X'} & V_{OZ}^{O'X'} & 0 \\ V_{OX}^{O'Y'} & V_{OY}^{O'Y'} & V_{OZ}^{O'Y'} & 0 \\ V_{OX}^{O'Z'} & V_{OY}^{O'Z'} & V_{OZ}^{O'Z'} & 0 \\ O'_X & O'_Y & O'_Z & 1 \end{pmatrix}.$$

Consider a more general case, presented in Fig. 2: two objects CS_L (X'Y'Z') and CS_L (X"Y"Z") are placed in the same space. Positions of these objects relative to the world coordinate system CS_W (XYZ) can be defined by:

$M_{OO'}$—transformation matrix from CS_W to CS_L of first object.
$M_{OO''}$—transformation matrix from CS_W to CS_L of second object.
$M_{O'O''}$—transformation matrix from CS_L of first object to CS_L of second object.

These matrices are linked by the following relationship [12]:

$$M_{OO''} = M_{O'O''} \cdot M_{OO'} \tag{1}$$

Thus, the mathematical algorithms used in computer graphics, allows us to determine the position of the target object in the world coordinate system, using the chain of matrix transformations to intermediate objects. Moreover, the chain of objects can be constructed in an arbitrary manner for the simplest calculations.

Application of the Transformation Matrix to the Bricard Linkage

The coordinate system of joint A as part of link AB $(X_B^{AB}, Y_B^{AB}, Z_B^{AB})$ is marked as M_A^{AB}. Similarly, the coordinate system of joint A as part of link AF is marked as M_A^{FA}.

The projection of axis X_A^{FA} to the axis X_A^{AB} can be marked as $\left(X_A^{FA} \rightarrow X_A^{AB}\right)$. In this case:

$\left(X_A^{FA} \rightarrow X_A^{AB}\right)$—projection of axis X_A^{FA} to the axis X_A^{AB};
$\left(X_A^{FA} \rightarrow Y_A^{AB}\right)$—projection of axis X_A^{FA} to the axis Y_A^{AB};
$\left(X_A^{FA} \rightarrow Z_A^{AB}\right)$—projection of axis X_A^{FA} to the axis Z_A^{AB}.

Similarly:

$\left(Y_A^{FA} \rightarrow X_A^{AB}\right)$—projection of axis Y_A^{FA} to the axis X_A^{AB};
$\left(Y_A^{FA} \rightarrow Y_A^{AB}\right)$—projection of axis Y_A^{FA} to the axis Y_A^{AB};
$\left(Y_A^{FA} \rightarrow Z_A^{AB}\right)$—projection of axis Y_A^{FA} to the axis Z_A^{AB};
$\left(Z_A^{FA} \rightarrow X_A^{AB}\right)$—projection of axis Z_A^{FA} to the axis X_A^{AB};
$\left(Z_A^{FA} \rightarrow Y_A^{AB}\right)$—projection of axis Z_A^{FA} to the axis Y_A^{AB};
$\left(Z_A^{FA} \rightarrow Z_A^{AB}\right)$—projection of axis Z_A^{FA} to the axis Z_A^{AB}.

Now it is possible to use these projections to obtain the transformation matrix from one coordinate system to another one:

$$\left(M_A^{AB} \rightarrow M_A^{FA}\right) = \begin{pmatrix} X_A^{FA} \rightarrow X_A^{AB} & X_A^{FA} \rightarrow Y_A^{AB} & X_A^{FA} \rightarrow Z_A^{AB} \\ Y_A^{FA} \rightarrow X_A^{AB} & Y_A^{FA} \rightarrow Y_A^{AB} & Y_A^{FA} \rightarrow Z_A^{AB} \\ Z_A^{FA} \rightarrow X_A^{AB} & Z_A^{FA} \rightarrow Y_A^{AB} & Z_A^{FA} \rightarrow Z_A^{AB} \end{pmatrix}. \quad (2)$$

Similarly, the transformation matrix from M_A^{FA} to M_E^{FA} can be defined as:

$$\left(M_A^{FA} \rightarrow M_E^{FA}\right) = \begin{pmatrix} X_E^{FA} \rightarrow X_A^{FA} & X_E^{FA} \rightarrow Y_A^{FA} & X_E^{FA} \rightarrow Z_A^{FA} \\ Y_E^{FA} \rightarrow X_A^{FA} & Y_E^{FA} \rightarrow Y_A^{FA} & Y_E^{FA} \rightarrow Z_A^{FA} \\ Z_E^{FA} \rightarrow X_A^{FA} & Z_E^{FA} \rightarrow Y_A^{FA} & Z_E^{FA} \rightarrow Z_A^{FA} \end{pmatrix}. \quad (3)$$

The transformation matrix from the coordinate system M_E^{FA} to the coordinate system M_A^{AB} can be obtained by sequential transformation from M_E^{FA} to M_A^{FA}, and from M_A^{FA} to M_A^{AB}. In this case:

$$\left(M_A^{AB} \rightarrow M_E^{FA}\right) = \left(M_A^{FA} \rightarrow M_E^{FA}\right) \cdot \left(M_A^{AB} \rightarrow M_A^{FA}\right). \quad (4)$$

Using (2) and (3) in (4) we finally obtain:

$$\left(M_A^{AB} \to M_E^{FA}\right) = \begin{pmatrix} X_A^{FA} \to X_A^{AB} & X_A^{FA} \to Y_A^{AB} & X_A^{FA} \to Z_A^{AB} \\ Y_A^{FA} \to X_A^{AB} & Y_A^{FA} \to Y_A^{AB} & Y_A^{FA} \to Z_A^{AB} \\ Z_A^{FA} \to X_A^{AB} & Z_A^{FA} \to Y_A^{AB} & Z_A^{FA} \to Z_A^{AB} \end{pmatrix}$$
$$\cdot \begin{pmatrix} X_E^{FA} \to X_A^{FA} & X_E^{FA} \to Y_A^{FA} & X_E^{FA} \to Z_A^{FA} \\ Y_E^{FA} \to X_A^{FA} & Y_E^{FA} \to Y_A^{FA} & Y_E^{FA} \to Z_A^{FA} \\ Z_E^{FA} \to X_A^{FA} & Z_E^{FA} \to Y_A^{FA} & Z_E^{FA} \to Z_A^{FA} \end{pmatrix} \tag{5}$$

The System of Equations for Bricard Linkage Joints

The matrix transformation principles showed above can be applied to each joint of Bricard linkage. The coordinate system of joint B as part of link AB is accepted as the fixed basic coordinate system. The coordinate system of joint D as part of link DE is accepted as the target coordinate system:

M_B^{AB}—the fixed basic coordinate system,
M_D^{DE}—the target coordinate system.

The target coordinate system can be transformed to the basic coordinate system by using two different contours shown in Fig. 4.

The transformation matrix from M_D^{DE} to M_B^{AB} by the chain of joints D, C, B:

$$\left(M_D^{DE} \to M_B^{AB}\right) = \left(M_D^{CD} \to M_D^{DE}\right) \cdot \left(M_C^{CD} \to M_C^{CD}\right) \cdot \left(M_C^{BC} \to M_C^{CD}\right)$$
$$\cdot \left(M_B^{BC} \to M_C^{BC}\right) \cdot \left(M_B^{AB} \to M_B^{BC}\right) \tag{6}$$

The transformation matrix from M_D^{DE} to M_B^{AB} by the chain of joints D, E, F, A, B:

$$\left(M_D^{DE} \to M_B^{AB}\right) = \left(M_E^{DE} \to M_D^{DE}\right) \cdot \left(M_E^{EF} \to M_E^{DE}\right) \cdot \left(M_F^{EF} \to M_E^{EF}\right)$$
$$\cdot \left(M_F^{FA} \to M_F^{EF}\right) \cdot \left(M_A^{FA} \to M_F^{FA}\right) \cdot \left(M_B^{AB} \to M_A^{FA}\right) \cdot \left(M_B^{AB} \to M_A^{AB}\right) \tag{7}$$

Fig. 4 Coordinate system
transformation contours

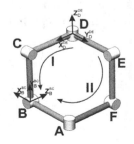

The transformation matrices for each joint of Bricard linkage are shown in Table 1. Considering (6) and (7):

Table 1 The transformation matrices of the Bricard linkage joints

Name	Coordinate systems	Transformation matrix
$\left(M_D^{CD} \rightarrow M_D^{DE}\right)$		$\begin{pmatrix} \cos(-\alpha_D) & \sin(-\alpha_D) & 0 \\ -\sin(\alpha_D) & \cos(-\alpha_D) & 0 \\ 0 & 0 & 1 \end{pmatrix}$
$\left(M_C^{CD} \rightarrow M_D^{CD}\right)$		$\begin{pmatrix} 0 & 0 & -1 \\ 0 & 1 & 0 \\ 1 & 0 & 0 \end{pmatrix}$
$\left(M_C^{BC} \rightarrow M_C^{CD}\right)$		$\begin{pmatrix} \cos(\alpha_C) & \sin(\alpha_C) & 0 \\ -\sin(\alpha_C) & \cos(\alpha_C) & 0 \\ 0 & 0 & 1 \end{pmatrix}$
$\left(M_B^{BC} \rightarrow M_C^{BC}\right)$		$\begin{pmatrix} 0 & 0 & -1 \\ 0 & 1 & 0 \\ 1 & 0 & 0 \end{pmatrix}$
$\left(M_B^{AB} \rightarrow M_B^{BC}\right)$		$\begin{pmatrix} \cos(\alpha_B) & \sin(\alpha_B) & 0 \\ -\sin(\alpha_B) & \cos(\alpha_B) & 0 \\ 0 & 0 & 1 \end{pmatrix}$
$\left(M_E^{DE} \rightarrow M_D^{DE}\right)$		$\begin{pmatrix} 0 & 0 & 1 \\ 0 & 1 & 0 \\ -1 & 0 & 0 \end{pmatrix}$
$\left(M_E^{EF} \rightarrow M_E^{DE}\right)$		$\begin{pmatrix} \cos(\alpha_E) & \sin(\alpha_E) & 0 \\ -\sin(\alpha_E) & \cos(\alpha_E) & 0 \\ 0 & 0 & 1 \end{pmatrix}$

(continued)

Table 1 (continued)

Name	Coordinate systems	Transformation matrix
$\left(M_F^{EF} \to M_E^{EF}\right)$		$\begin{pmatrix} 0 & 0 & 1 \\ 0 & 1 & 0 \\ -1 & 0 & 0 \end{pmatrix}$
$\left(M_F^{FA} \to M_F^{EF}\right)$		$\begin{pmatrix} \cos(-\alpha_F) & \sin(-\alpha_F) & 0 \\ -\sin(-\alpha_F) & \cos(-\alpha_F) & 0 \\ 0 & 0 & 1 \end{pmatrix}$
$\left(M_A^{FA} \to M_F^{FA}\right)$		$\begin{pmatrix} 0 & 0 & -1 \\ 0 & 1 & 0 \\ 1 & 0 & 0 \end{pmatrix}$
$\left(M_A^{AB} \to M_A^{FA}\right)$		$\begin{pmatrix} \cos(\alpha_A) & \sin(\alpha_A) & 0 \\ -\sin(\alpha_A) & \cos(\alpha_A) & 0 \\ 0 & 0 & 1 \end{pmatrix}$
$\left(M_B^{AB} \to M_A^{AB}\right)$		$\begin{pmatrix} 0 & 0 & 1 \\ 0 & 1 & 0 \\ -1 & 0 & 0 \end{pmatrix}$

$$
\begin{aligned}
\left(M_D^{DE} \to M_B^{AB}\right) = &\begin{pmatrix} \cos(-\alpha_D) & \sin(-\alpha_D) & 0 \\ -\sin(\alpha_D) & \cos(-\alpha_D) & 0 \\ 0 & 0 & 1 \end{pmatrix} \cdot \begin{pmatrix} 0 & 0 & -1 \\ 0 & 1 & 0 \\ 1 & 0 & 0 \end{pmatrix} \\
&\cdot \begin{pmatrix} \cos(\alpha_C) & \sin(\alpha_C) & 0 \\ -\sin(\alpha_C) & \cos(\alpha_C) & 0 \\ 0 & 0 & 1 \end{pmatrix} \cdot \begin{pmatrix} 0 & 0 & -1 \\ 0 & 1 & 0 \\ 1 & 0 & 0 \end{pmatrix} \\
&\cdot \begin{pmatrix} \cos(\alpha_B) & \sin(\alpha_B) & 0 \\ -\sin(\alpha_B) & \cos(\alpha_B) & 0 \\ 0 & 0 & 1 \end{pmatrix}
\end{aligned} \tag{8}
$$

$$\left(M_D^{DE} \rightarrow M_B^{AB} \right) = \begin{pmatrix} N_{XX}^{DCB} & N_{XY}^{DCB} & N_{XZ}^{DCB} \\ N_{YX}^{DCB} & N_{YY}^{DCB} & N_{YZ}^{DCB} \\ N_{ZX}^{DCB} & N_{ZY}^{DCB} & N_{ZZ}^{DCB} \end{pmatrix} \tag{9}$$

where:

$$N_{XX}^{DCB} = \cos(\alpha_C) \cdot \sin(\alpha_B) \cdot \sin(\alpha_D) - \cos(\alpha_B) \cdot \cos(\alpha_D); \tag{10}$$

$$N_{XY}^{DCB} = -\cos(\alpha_D) \cdot \sin(\alpha_B) - \cos(\alpha_B) \cdot \cos(\alpha_C) \cdot \sin(\alpha_D); \tag{11}$$

$$N_{XZ}^{DCB} = -\sin(\alpha_D) \cdot \sin(\alpha_C); \tag{12}$$

$$N_{YX}^{DCB} = -\cos(\alpha_B) \cdot \sin(\alpha_D) - \cos(\alpha_D) \cdot \cos(\alpha_C) \cdot \sin(\alpha_B); \tag{13}$$

$$N_{YY}^{DCB} = \cos(\alpha_B) \cdot \cos(\alpha_D) \cdot \cos(\alpha_C) - \sin(\alpha_B) \cdot \sin(\alpha_D); \tag{14}$$

$$N_{YZ}^{DCB} = \cos(\alpha_D) \cdot \sin(\alpha_C); \tag{15}$$

$$N_{ZX}^{DCB} = -\sin(\alpha_B) \cdot \sin(\alpha_C); \tag{16}$$

$$N_{ZY}^{DCB} = \cos(\alpha_B) \cdot \sin(\alpha_C); \tag{17}$$

$$N_{ZZ}^{DCB} = -\cos(\alpha_C). \tag{18}$$

And matrices obtained by another one contour:

$$\left(M_D^{DE} \rightarrow M_B^{AB} \right) = \begin{pmatrix} 0 & 0 & 1 \\ 0 & 1 & 0 \\ -1 & 0 & 0 \end{pmatrix} \cdot \begin{pmatrix} \cos(\alpha_E) & \sin(\alpha_E) & 0 \\ -\sin(\alpha_E) & \cos(\alpha_E) & 0 \\ 0 & 0 & 1 \end{pmatrix} \cdot$$

$$\cdot \begin{pmatrix} 0 & 0 & 1 \\ 0 & 1 & 0 \\ -1 & 0 & 0 \end{pmatrix} \cdot \begin{pmatrix} \cos(-\alpha_F) & \sin(-\alpha_F) & 0 \\ -\sin(-\alpha_F) & \cos(-\alpha_F) & 0 \\ 0 & 0 & 1 \end{pmatrix} \cdot \begin{pmatrix} 0 & 0 & -1 \\ 0 & 1 & 0 \\ 1 & 0 & 0 \end{pmatrix}$$

$$\cdot \begin{pmatrix} \cos(\alpha_A) & \sin(\alpha_A) & 0 \\ -\sin(\alpha_A) & \cos(\alpha_A) & 0 \\ 0 & 0 & 1 \end{pmatrix} \cdot \begin{pmatrix} 0 & 0 & -1 \\ 0 & 1 & 0 \\ 1 & 0 & 0 \end{pmatrix}$$

$$\tag{19}$$

$$\left(M_D^{DE} \rightarrow M_B^{AB} \right) = \begin{pmatrix} N_{XX}^{DEFAB} & N_{XY}^{DEFAB} & N_{XZ}^{DEFAB} \\ N_{YX}^{DEFAB} & N_{YY}^{DEFAB} & N_{YZ}^{DEFAB} \\ N_{ZX}^{DEFAB} & N_{ZY}^{DEFAB} & N_{ZZ}^{DEFAB} \end{pmatrix} \tag{20}$$

where

$$N_{XX}^{DEFAB} = -\cos(\alpha_F);\tag{21}$$

$$N_{XY}^{DEFAB} = \cos(\alpha_A) \cdot \sin(\alpha_F);\tag{22}$$

$$N_{XZ}^{DEFAB} = -\sin(\alpha_F) \cdot \sin(\alpha_A);\tag{23}$$

$$N_{YX}^{DEFAB} = \cos(\alpha_E) \cdot \sin(\alpha_F);\tag{24}$$

$$N_{YY}^{DEFAB} = \cos(\alpha_F) \cdot \cos(\alpha_A) \cdot \cos(\alpha_E) - \sin(\alpha_A) \cdot \sin(\alpha_E);\tag{25}$$

$$N_{YZ}^{DEFAB} = -\cos(\alpha_A) \cdot \sin(\alpha_E) - \cos(\alpha_F) \cdot \cos(\alpha_E) \cdot \sin(\alpha_A);\tag{26}$$

$$N_{ZX}^{DEFAB} = -\sin(\alpha_F) \cdot \sin(\alpha_E);\tag{27}$$

$$N_{ZY}^{DEFAB} = -\cos(\alpha_E) \cdot \sin(\alpha_A) - \cos(\alpha_F) \cdot \cos(\alpha_A) \cdot \sin(\alpha_E);\tag{28}$$

$$N_{ZZ}^{DCB} = \cos(\alpha_F) \cdot \sin(\alpha_A) \cdot \sin(\alpha_E) - \cos(\alpha_A) \cdot \cos(\alpha_E).\tag{29}$$

Bricard linkage is a closed mechanism, so both of these chains must be identical. Thus, each element of the chain must be identical to the element of the second chain. In this case:

$$\begin{cases} N_{XX}^{DCB} = N_{XX}^{DEFAB}, \\ N_{XY}^{DCB} = N_{XY}^{DEFAB}, \\ N_{XZ}^{DCB} = N_{XZ}^{DEFAB}, \\ N_{YX}^{DCB} = N_{YX}^{DEFAB}, \\ N_{YY}^{DCB} = N_{YY}^{DEFAB}, \\ N_{YZ}^{DCB} = N_{YZ}^{DEFAB}, \\ N_{ZX}^{DCB} = N_{ZX}^{DEFAB}, \\ N_{ZY}^{DCB} = N_{ZY}^{DEFAB}, \\ N_{ZZ}^{DCB} = N_{ZZ}^{DEFAB}. \end{cases}\tag{30}$$

Using (10)–(18) and (21)–(29) in (30):

$$\begin{cases} \cos(\alpha_C) \cdot \sin(\alpha_B) \cdot \sin(\alpha_D) - \cos(\alpha_B) \cdot \cos(\alpha_D) = -\cos(\alpha_F), \\ -\cos(\alpha_D) \cdot \sin(\alpha_B) - \cos(\alpha_B) \cdot \cos(\alpha_C) \cdot \sin(\alpha_D) = \cos(\alpha_A) \cdot \sin(\alpha_F), \\ -\sin(\alpha_D) \cdot \sin(\alpha_C) = -\sin(\alpha_F) \cdot \sin(\alpha_A), \\ -\cos(\alpha_B) \cdot \sin(\alpha_D) - \cos(\alpha_D) \cdot \cos(\alpha_C) \cdot \sin(\alpha_B) = \cos(\alpha_E) \cdot \sin(\alpha_F), \\ \cos(\alpha_B) \cdot \cos(\alpha_D) \cdot \cos(\alpha_C) - \sin(\alpha_B) \cdot \sin(\alpha_D) = \cos(\alpha_F) \cdot \cos(\alpha_A) \cdot \cos(\alpha_E) - \sin(\alpha_A) \cdot \sin(\alpha_E), \\ \cos(\alpha_D) \cdot \sin(\alpha_C) = -\cos(\alpha_A) \cdot \sin(\alpha_E) - \cos(\alpha_F) \cdot \cos(\alpha_E) \cdot \sin(\alpha_A), \\ -\sin(\alpha_B) \cdot \sin(\alpha_C) = -\sin(\alpha_F) \cdot \sin(\alpha_E), \\ \cos(\alpha_B) \cdot \sin(\alpha_C) = -\cos(\alpha_E) \cdot \sin(\alpha_A) - \cos(\alpha_F) \cdot \cos(\alpha_A) \cdot \sin(\alpha_E), \\ -\cos(\alpha_C) = \cos(\alpha_F) \cdot \sin(\alpha_A) \cdot \sin(\alpha_E) - \cos(\alpha_A) \cdot \cos(\alpha_E). \end{cases} \tag{31}$$

The Maple system of symbolic calculations developed by Waterloo Maple Inc., allows one to simplify the mathematical calculations [13]. Here are formulas of joints rotation angles, calculated by Maple for (31):

$$\alpha_A = \pi - \arccos\left(\frac{\cos(\alpha_B)}{\cos(\alpha_B) + 1}\right); \tag{32}$$

$$\alpha_C = \pi - \arccos\left(\frac{\cos(\alpha_B)}{\cos(\alpha_B) + 1}\right); \tag{33}$$

$$\alpha_D = \alpha_B; \tag{34}$$

$$\alpha_E = \pi - \arccos\left(\frac{\cos(\alpha_B)}{\cos(\alpha_B) + 1}\right); \tag{35}$$

$$\alpha_F = \alpha_B \tag{36}$$

Equations (32)–(36) gives the range of rotation angles for the Bricard linkage joints. Each link rotates in the range $[-120; 120]$.

Angular velocities of joints A, C, E can be obtained by taking the first derivative of (32), (33) and (35):

$$\omega = -\frac{\left(\dfrac{\sin(\alpha_B)}{\cos(\alpha_B) + 1} - \dfrac{\cos(\alpha_B) \cdot \sin(\alpha_B)}{(\cos(\alpha_B) + 1)^2}\right)}{\sqrt{\dfrac{1 - \cos(\alpha_B)^2}{(\cos(\alpha_B) + 1)^2}}} \tag{37}$$

The first derivative of (37) allows us to obtain the angular acceleration (a formula is given in the form, adopted for using in computer systems):

$$\varepsilon = (\cos(x)^2/(\cos(x)+1)^2 - (2 \cdot \sin(x)^2/(\cos(x)+1)^2 - \cos(x)/(\cos(x)+1)$$
$$+ (2 \cdot \cos(x) \cdot \sin(x)^2)/(\cos(x)+1)^3)/(1 - \cos(x)^2/(\cos(x)+1)^2)^{1/2}$$
$$- ((\sin(x)/(\cos(x)+1) - (\cos(x) \cdot \sin(x))/(\cos(x)+1)^2)$$
$$\cdot ((2 \cdot \cos(x)^2 \cdot \sin(x))/(\cos(x)+1)^3 - (2 \cdot \cos(x)$$
$$\cdot \sin(x))/(\cos(x)+1)^2))/(2 \cdot (1 - \cos(x)^2/(\cos(x)+1)^2)^{3/2}))$$

Conclusion

So, the system of equations necessary to make kinematic research of Bricard linkage is obtained in this article. This allows us to the relationship between rotation angles of Bricard mechanism links. As a result, this research gives formulas of angular velocity and acceleration of Bricard linkage joints.

Two Bricard linkage joints (D, F) rotates the same as driving joint (B). Another three joints (A, C, E) rotates by another way, but identical with each over. Each link of Bricard mechanism rotates at range [−120; 120].

Transformation matrices can be used to determine the kinematic parameters of Bricard linkage. The advantage of the proposed method is its universality. This method can be used to find kinematic parameters of any modifications of the Bricard linkage. Furthermore, the proposed method is fully formalized and can be used to develop a computer program analyzing the kinematic parameters of the Bricard linkage.

References

1. Bricard R (1927) Lectures on kinematics, vol 2. Paris (in French)
2. Kozlovsky MZ, Evgrafov AN, Semenov YA, Slouch AV (2008) Theory of mechanisms and machines. Publishing Center "Akademia", Moscow. 560 p. (in Russian)
3. Dimentberg FM (1982) Theory of spatial hinged mechanisms. Nauka, Moscow (in Russian)
4. Zinoviev VA (1952) Spatial mechanisms with lower kinematic pairs. Gostehizdat, Moscow (in Russian)
5. Mersalov NI (1951) Theory of spatial mechanisms. Gostehizdat, Moscow (in Russian)
6. Mudrov PG (1976) Spatial linkages with rotational kinematic pairs. Izd-vo Kazan. un-ta, Kazan (in Russian)
7. Yarullin MG, Galiullin IA (2012) Analytical review of the 6R linkages researches In: Analytical mechanics, stability and control: proceedings of 10th international scientific conference named after Chetaev. Izd-vo Kazan. un-ta, Kazan. pp 117–126 (in Russian)
8. Chen Y (2003) Design of structural mechanisms. Dissertation. St Hugh's College, University of Oxford, Oxford
9. Racila L, Dahan M (2007) Bricard mechanism used as translator. IFToMM: proceedings of twelfth world congress in mechanism and machine science. Besancon, France, pp 337–341

10. Racila L, Dahan M (2011) 6R parallel translational device. Linkages and cams: proceedings of 13th world congress in mechanism and machine science. Guanajuato, Mexico, pp 85–93
11. Yaozhi L, Ying Y, Jingjing L (2008) A retractable structure based on Bricard linkages and rotating rings of tetrahedral. Int J Solids Struct 45:620–630
12. Yarullin MG, Galiullin IA (2013) Synthesis of mobile mechanisms In: Problems and perspectives of aviation, land transport and energy: proceedings of 6th international conference ANTE-2013. Izd-vo Kazan. un-ta, Kazan, pp 23–31 (in Russian)
13. Char B, Fee G, Geddes K, Gonnet G, Monagan M (1986) A tutorial introduction to Maple. J Symbolic Comput 2:179–200

Drive Selection of Multidirectional Mechanism with Excess Inputs

Alexander N. Evgrafov and Gennady N. Petrov

Abstract Mechanisms with excess inputs are mechanisms in which the number of engines is higher than the number of degrees of freedom. While the mechanism is in operation, some engines are disengaging, while other engines are switching on, whereas the number of active engines at each given moment equals the number of degrees of freedom. These mechanisms create better conditions for power transmission than mechanisms without excess inputs. However to realize full potential of these mechanisms it is necessary to solve the problem of operational drive selection. This article describes a mechanism of a spatial platform with excess inputs. For the purpose of this article, the following criteria were used to carry out a quality assessment of the configuration presented below: Jacobian determinant of equations in sequence for geometric analysis of the mechanism, sum of squares of the balancing forces and minimal natural frequency of the mechanism with fixed engines and flexible transmission device. This article also considers variations of motion of the movable operating element at pre-set conditions with an alternative engines activation under all three mentioned criteria, as well as variations for simultaneous operation of all engines.

Keywords Spatial mechanism · Multidirectional mechanism · Excess input · Drive selection

Introduction

Closed linkwork mechanisms are complex mechanical systems, and are widely used in mechanical engineering. Position functions of such mechanisms are non-linear and link output coordinates, defining positions of movable operating elements, with

A.N. Evgrafov (✉) · G.N. Petrov
Peter the Great Saint-Petersburg Polytechnic University, Saint Petersburg, Russia
e-mail: a.evgrafov@spbstu.ru

G.N. Petrov
e-mail: gnpet@mail.ru

© Springer International Publishing Switzerland 2016
A. Evgrafov (ed.), *Advances in Mechanical Engineering*,
Lecture Notes in Mechanical Engineering, DOI 10.1007/978-3-319-29579-4_4

input coordinates, setting positions of drive output links [1–6]. Yet a function of multidirectional mechanism position, which realizes its programmed motion through coordinated controlled motion of several engines, is even more complex [7–9]. In most cases, these nonlinear functions of position are multivalued. The same set of input coordinates corresponds to several positions (configurations) of the system.

Problem Statement

Multivaluedness of the above-mentioned function leads to occurrence of specific singular positions, which are bifurcation points in the area of system configuration. In those specific positions, mechanism acquires an additional degree of freedom, while simultaneously reaction forces in kinematic pairs increase without bound. It results in creation of "self-locking areas" in a vicinity of the specific positions, given the Coulomb friction force. Thereby problems of motion selection continuously arise. In the majority of cases those problems relate to positioning—repositioning the system from the given initial position to the pre-set target position. At the same time the need to ensure the choice of "best" sequence of configurations, while controlling the engines, arises. Systematic selection of the optimal path or the choice of optimal laws of motion along the trajectory could be carried through this present sequence of configurations. There is also a possibility to have trajectory and the law of motion pre-defined, thus, in such a case, it is only necessary to determine the optimal calculus of variation for driving forces and momentum.

Thus, we need the solution of practical problems resulting in the need of initially solving theoretical problems of quality assessment for various mechanism configurations with various trajectories. Criteria of quality configuration may be selected in various ways. This article discusses the criteria of quality, reflecting the system's degree of proximity to specific positions. Some of these criteria were considered in the previous research [10], for instance, contact angle [11]. However, this criterion proved ineffective in movement optimization of a mechanism with multiple degrees of freedom.

Formation of other criteria could be based on the following properties of the specific positions:

1. The Jacobian determinant of equations in a sequence for geometric analysis of a mechanism is equal to zero, while in a specific position. Application of this criterion is discussed more in the previous research [12].
2. Driving forces and momentum, counterbalancing ultimate external load constant in magnitude, increase without bound in the vicinity of specific positions. Application of this criterion is discussed more in the previous research [13].
3. In case some additional local degree of freedom appears at the specific position, a mechanism's natural frequency becomes zero. A detailed description of this criterion is discussed more in the previous research [14].

Theory

Figure 1 presents a mechanism—a platform where the desktop is set to eight legs. Platform legs are hydraulic cylinders, connected by spherical hinges with the platform itself and its base. The mechanism has six degrees of freedom and eight inputs. For standard operation it is sufficient for the mechanism to have six inputs, thus engines may operate in an alternative mode of activation, where at any given moment only six of eight engines (two engines are turned off), or all engines at once, are connected with the mechanism.

To determine which of the engines to activate at any given moment when of its position on the platform, it is suggested to introduce three criteria for the quality of the platform configuration (these are based on three properties of the specific positions described above). To gain deeper insight on the subject it is recommended to review the inverse problem of spatial platform geometric analysis.

We consider the position of point K of the platform (x_K, y_K, z_K) and Euler angles (ψ, θ, ϕ) that define the orientation of the moving coordinate system x', y', z' in relation to the fixed coordinate system x, y, z as given values. We assume that leg drives with lengths $\ell_i, i = 1, \ldots, 6$ are activated at the current moment. We introduce matrix columns $\eta = (x_K; y_K; z_K; \psi; \theta; \phi)^T$ and $\ell = (\ell_1; \ldots; \ell_6)^T$, so that the calculation of the lengths based on the known column η becomes a simple task:

$$\ell_i = \|\mathbf{R}_i(\eta) - \mathbf{R}_{0i}\|, \quad i = 1, \ldots, 6$$

where $\mathbf{R}_i(\eta)$—radius-vector of fixed point of attachment of the i leg to the platform, and $\mathbf{R}_{0i}$—radius-vector of fixed point of attachment of the i leg to the pillar.

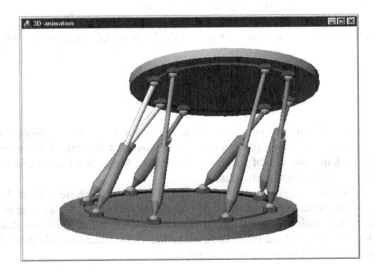

Fig. 1 The spatial platform

The direct problem of geometric analysis is more complex. Defining a column η by known lengths $\ell = (\ell_1; \ldots; \ell_6)^T$ requires a solution for the system of six equations with six unknowns:

$$\Phi_i(\eta, \ell_i) = \|\mathbf{R}_i(\eta) - \mathbf{R}_{0i}\| - \ell_i = 0, \quad i = 1, \ldots, 6.$$

We may define the Jacobian of this system of equations as follows:

$$J = \det\left(\frac{\partial \Phi}{\partial \eta}\right), \quad \text{where } \Phi = (\Phi_1; \ldots; \Phi_6)^T, \frac{\partial \Phi}{\partial \eta} - \text{Jacobian matrix.}$$

It is known that Jacobian equals to zero, while in specific position. However, it is possible to avoid the vicinity of specific positions by alternating inputs. Thus, we may define a Jacobian module of equations in sequence for geometric analysis as the first criterion of quality of the mechanism configuration.

We may present the angular velocity of the platform in the following form:

$$\boldsymbol{\omega} = \dot{\psi} \cdot \mathbf{k} + \dot{\theta} \cdot \mathbf{n} + \dot{\phi} \cdot \mathbf{k}',$$

where $\mathbf{k}, \mathbf{n}, \mathbf{k}'$—corresponding unit vectors.

We assume that the force $\mathbf{F}$ and momentum $\mathbf{M}$ are applied to the platform at the point K.

Let us introduce matrix-column $P = (F_x; F_y; F_z; M_1; M_2; M_3)^T$, where F_x, F_y, F_z—projections of the vector $\mathbf{F}$ on the fixed coordinate system axis; M_1, M_2, M_3—projections of the vector $\mathbf{M}$ in the direction of the unit vectors $\mathbf{k}, \mathbf{n}, \mathbf{k}'$:

$$\mathbf{M} = M_1 \cdot \mathbf{k} + M_2 \cdot \mathbf{n} + M_3 \cdot \mathbf{k}'.$$

Evidently, drive forces $F_D = (F_{D1}; \ldots; F_{D6})^T$ may be determined from the formula

$$F_D = -\left(\frac{\partial \eta}{\partial \ell}\right)^T \cdot P.$$

When in the vicinity of specific positions of the mechanism, the absolute value of at least one of the driving forces increases without bound. Thus, it is considered reasonable to define the sum of squares of the driving forces as a second criterion of quality of the mechanism configuration.

To gain deeper insight on the subject we consider a machine with a flexible mechanism, taking into consideration stiffness of the leg drives. We assume that six inputs are active while two are disabled (thus we are not taking into consideration flexibility properties of those two). We assume that the mechanism performs natural oscillations in a vicinity of the specific position.

We introduce matrix columns: $\theta = (\theta_1; \ldots; \theta_6)^\mathrm{T}$—column, representing a small strain of resilient members and $\rho_S = (\Delta x_S; \Delta y_S; \Delta z_S; \Delta\phi_{x'}; \Delta\phi_{y'}; \Delta\phi_{z'})^\mathrm{T}$—column, representing platform deviations from the equilibrium position, where $\Delta x_S, \Delta y_S, \Delta z_S$—deviation of the Platform's center of mass towards corresponding axes; $\Delta\phi_{x'}, \Delta\phi_{y'}, \Delta\phi_{z'}$—deviation of the Platform's alignment towards corresponding axes of the moving coordinate system x', y', z'.

Based on the previous work done by Loitsiansky and Lurie [15] we know the relation of deviations of the platform's orientation to deviations of Euler angles:

$$\Delta\phi_{x'} = \Delta\psi \sin\theta \sin\phi + \Delta\theta \cos\phi,$$
$$\Delta\phi_{y'} = \Delta\psi \sin\theta \cos\phi - \Delta\theta \sin\phi,$$
$$\Delta\phi_{z'} = \Delta\psi \cos\theta + \Delta\phi.$$

We introduce the square diagonal matrix:

$$A = diag\left(m; m; m; I_{x'}; I_{y'}; I_{z'}\right), \quad C = diag(C_1; \ldots; C_6),$$

where m—mass of platform, $I_{x'}, I_{y'}, I_{z'}$—primary central axial moments of inertia, and $C_1, \ldots, C_6$—stiffness of the active drives.

For the purpose of this article, we disregard mass of platform legs, thus we may present the equation of the resilient members' natural oscillations in the following form

$$\left(\frac{\partial\rho_S}{\partial\ell}\right)^\mathrm{T} A \frac{\partial\rho_S}{\partial\ell}\ddot{\theta} + C\theta = 0.$$

Accordingly, we introduce symbol $H = C^{-1}\left(\frac{\partial\rho_S}{\partial\ell}\right)^\mathrm{T} A \frac{\partial\rho_S}{\partial\ell}$.

The natural frequencies of a flexible mechanism are equal to reciprocals of the matrix H's eigenvalue roots. One of the natural frequencies tends to zero, once the mechanism is in a vicinity of the specific position.

In consequence, we may define the value of the fundamental frequency of a flexible mechanism as the third criterion of quality of the mechanism configuration. However, it should be emphasized that, the mechanism flexibility is introduced conditionally, for the purpose of this work, while mechanism's natural frequencies are defined within a range of static positions along the pre-set trajectory.

In case of simultaneous operation of all engines, the selection of driving forces on pre-determined movement and at a given load becomes ambiguous. As a result, we may outline a system of six equations with eight unknowns and an infinite set of solutions, which in turn allows us to set the problem to minimize the sum of squares of the driving forces. Detailed description of the solution is discussed more in the previous research [14].

The analytical solution of the problem is rather laborious, thus, the authors decided to use the numerical calculation method. Typically software tools such as

Fig. 2 Diagrams represent sum of squares of the platform's driving forces, given that six engines are operating according to the criterion of minimum sum of squares of the driving forces (*a*) and at simultaneous operation of all eight engines (*b*)

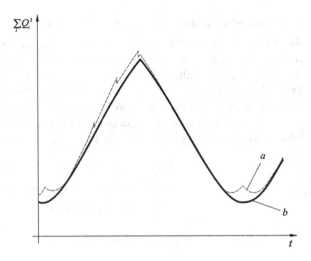

Mathcad and Matlab are used in such cases (Both Mathcad and Matlab usage is discussed in the previous research: [16–18] and [19] respectively). For convenience, obtained results are visualized through animation of mechanisms kinematics (as discussed in the previous research [20]). However, for the purpose of this work, and to be able to obtain an interactive visualization of the calculation results of a complex spatial mechanism, authors selected "Model Vision" software (discussed in the previous research [21]).

Figure 2 shows two diagrams, representing the sum of squares of the platform's driving forces, given that six engines are operating according to the criterion of minimum sum of squares of the driving forces (a) and at simultaneous operation of all eight engines (b).

Conclusion

The resulting driving forces provide a minimum sum of squares of the driving forces at any given point of the trajectory.

It is possible to evaluate the energy efficiency of the operational mode where all eight engines function simultaneously by comparing the sum of squares of driving forces with the minimum value of such a sum attained when only six engines are functioning. Taking into consideration criterion of the sum of squares of the driving forces and criterion of the fundamental frequency, authors may conclude that the operational mode where all eight engines function simultaneously is considered more beneficial. Additionally, ease of management may be considered as an advantage of the alternative inputs selection.

References

1. Dimentberg FM (1982) The theory of spatial hinge mechanisms. Moscow, Nauka, 336 p. (In Russian)
2. Zinoviev VA (1952) Spatial mechanisms with lower pairs. Moscow, Gostehizdat, 432 p. (In Russian)
3. Mertsalov NI (1951) The theory of spatial mechanisms. Moscow, Gostehizdat, 206 p. (In Russian)
4. Uicker JJ (1963) Velocity and acceleration analysis of spatial mechanisms using 4 × 4 matrices. Northwestern University, The Technological Institute, Evanston, Illinois
5. Mudrov PG (1976) Spatial mechanisms with rotational pairs. Publishing House of Kazan University, Kazan, 264 p. (In Russian)
6. Hrostitsky AA, Evgrafov AN, Tereshin VA (2011) The geometry and kinematics of spatial six-bar mechanisms with excess links. In: Scientific and technical news of SPbPU. Publishing House of the Polytechnic University, St-Petersburg, vol 2 (123), pp 170–176. (In Russian)
7. Kolovsky MZ, Evgrafov AN, Semenov YA, Slousch AV (2000) Advanced theory of mechanisms and machines. Springer, Berlin, 394 p
8. Kolovsky MZ, Evgrafov AN, Semenov YA, Slousch AV (2013) Theory of mechanisms and machines: textbook for high school students, 4th revised edn. Publishing Center, Moscow, Akademia, 560 p. (In Russian)
9. Evgrafov AN (2015) Theory of mechanisms and machines: textbook/A.N. Evgrafov, M.Z. Kolovsky, G.N. Petrov. Publishing House of the Polytechnic University, St.-Petersburg, 248 p. (In Russian)
10. Evgrafov AN, Petrov GN Quality criteria used in computer aided mechanism synthesis. In: 2nd international scientific and practical conference "modern engineering: science and education". Publishing House of the Polytechnic University, St.-Petersburg, pp 48–52. (In Russian)
11. Evgrafov AN (1997) Contact angle as a quality criteria of the force transmission. Mashinovedenie. Collection of scientific papers. Publishing house of the Polytechnic University, St.-Petersburg, pp 84–90. (In Russian)
12. Kolovsky MZ (1997) Quality criterion of multidirectional lever mechanisms positions, no 2. (In Russian)
13. Kolovsky MZ, Slousch AV (1998) Movement control of the closed lever mechanisms with excess inputs. Mat. of XXV summer school "Nonlinear oscillations–97". St.-Petersburg Institute of Problems of Mechanical Engineering of RAS. (In Russian)
14. Kolovsky MZ, Petrov GN, Slousch AV (2000) Movement control of the closed lever mechanisms with several degrees of freedom. J Mach Manuf Reliab, no 4. (In Russian)
15. Loitsiansky LG, Lurie AI (1982) Theoretical mechanics course: in 2 volumes, vol I. Statics and kinematics. Nauka, Moscow, 352 p. (In Russian)
16. Petrov GN (1993) Computer aided kinetostatic algorithm for closed lever mechanisms. J Mach Manuf Reliab (3). (In Russian)
17. Evgrafov AN, Petrov GN Geometric and kinetostatic analysis of plain lever mechanisms of the second class. Theor Mech Mach 2:50–63. (In Russian)
18. Ziborov KA, Matsyuk IN, Shlyahov EM (2010) Power analysis of mechanisms using Mathcad. Theor Mech Mach 8(1):83–88. (in Russian)
19. Mkrtychev OV (2013) Computer simulations in the plain mechanisms power calculation. Theor Mech Mach 11(1):77–83. (In Russian)
20. Evgrafov AN, Petrov GN (2008) Computer animation of kinematic schemes in excel and mathcad. Theor Mech Mach 6(1):71–80. (In Russian)
21. Petrov GN (2004) Computer simulation of mechanical systems in the "ModelVision". Theor Mech Mach 2(1):75–79. (In Russian)

Engineering Calculations of Bolt Connections

Alexander A. Sukhanov

Abstract The paper deals with the theory and practice of calculations of bolt connections. This focuses on the engineering approach to obtaining convenient approximate analytical formulas for the calculation of strong bolted joints. Explicit expressions for these stresses have been obtained and proved to be in good agreement with existing engineering tables and numerical simulations.

Keywords Bolt · Bolt connection · Stress · Calculations of strength · Engineering formulas

Introduction

The existing literature on bolted joints (see e.g. [1–3]) has proved to be sufficiently deep for theoretical research on calculation of the forces and stresses in bolted joints. But little attention has been given to the practical use of the results. The reference literature (see e.g. [4]) on the contrary, provides practical tables for particular bolts, depending on the grade of steel bolt material, but there is no analytical dependence allowing us to expand the table for other parameter values (friction, torque etc.). This paper fills a gap between the theory and practice of bolting. The classification of bolting and the conclusion are simple and convenient engineering formulas for calculation of forces, stress, torque, safety factors.

A.A. Sukhanov (✉)
Peter the Great Saint-Petersburg Polytechnic University, Saint Petersburg, Russia
e-mail: Alexeevich@post.ru

© Springer International Publishing Switzerland 2016
A. Evgrafov (ed.), *Advances in Mechanical Engineering*,
Lecture Notes in Mechanical Engineering, DOI 10.1007/978-3-319-29579-4_5

Fig. 1 Bolt connection

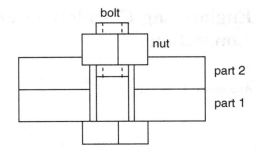

Purpose and Modes of Bolt Connections

Bolted joints are designed to reliably connect two or more parts with a bolt, a nut and possibly conventional and/or grower-plates (Fig. 1). It is also possible that a special case of a screw connection, when the bolt is screwed into one of the parts, is essentially a nut.

There are basically three different modes of bolt connections.

Bayonet Connection Mode

In the pin connection a bolt locks securely connected parts, impeding their cross (tangential) offset by the forces Q. When this bolt is installed in the hole (usually from under the scan) it is without play or even with a slight interference fit (Fig. 2). Commonly such bolts, called fitter bolts, are used as shear keys.

In order not to overload the additional bolt stresses in a pin mode, a backing nut is twisted into a little torque M_3^{min}, creating a small initial axial tensile force F_0^{min} and shear stress τ^{min}.

Mode of Longitudinal Confinement

In the longitudinal retaining bolt and nut, securely fixed parts are connected, preventing their longitudinal (normal) disengage under the action of N (Fig. 3).

To ensure the density interface (lack of separation of parts) the nut is tightened with a sufficiently large force optimal torque M_3^{opt}. This bolt experiences average axial tension F_0^{opt} and shear stress τ^{opt}.

Fig. 2 Pin joint

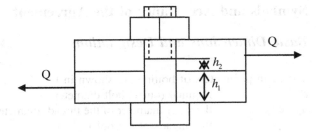

Fig. 3 Longitudinal hold

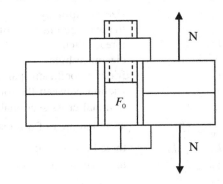

Fig. 4 Tight connection

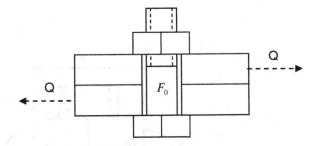

Mode Tight Connection

In a tight connection, a bolt and a nut securely fix the parts to prevent their possible cross (tangential) shifting under the influence of possible transverse shear forces Q (Fig. 4).

In this mode, the bolt stress tests only preliminary tightening because transverse shear forces Q are fully compensated by friction forces. Therefore, to ensure that a maximum density connection bolt is tightened, the maximum torque is set to M_3^{max}. This bolt undergoes maximum axial tensile F_0^{max} and shear stress τ^{max}.

Symbols and Acceptance of the Agreement

Basic Dimensions and Designations

The main parameters of bolting are shown in Fig. 5.

d	nominal (outer) bolt diameter
d_0	the inner diameter of the thread (diameter of the solid body below the threads of the bolt)
$d_P = \frac{d_0 + d}{2}$	the average diameter of the thread
d_K	size nut turnkey
d_Γ	the average diameter of the support ring contact nut
h	thread pitch
δ	clearance hole
f	friction coefficient pairs bolt/nut, nut/washer
$f_Д$	friction between the joined parts
$S = \frac{\pi d^2}{4}$	nominal cross-sectional area of the bolt
$S_0 = \frac{\pi d_0^2}{4}$	sectional area of a threaded bolt
$k_0 = \frac{d_0}{d}, k_P = \frac{d_P}{d}$	thread coefficients
$k_\Gamma = \frac{d_\Gamma}{d}$	nut coefficient
M_3	nut torque

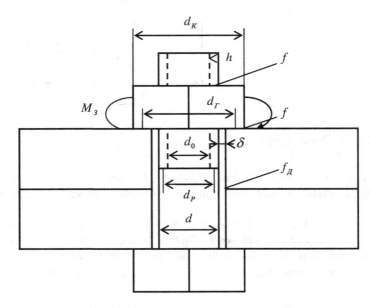

Fig. 5 The main parameters of bolting

Accepted Agreement

1. The coefficient of friction f is largely dependent on the state of the contacting surfaces, in particular on the purity and availability of processing lubricant. For steel parts

$$f = 0.12 - 0.15 - 0.18,$$

which corresponds to the lower value of the lubricated surfaces, and the upper— roughly treated dry [1]. As the medium and most commonly used values of the coefficient of friction they will take

$$f = 0.15, \quad f_{\varPi} = 0.18. \tag{1}$$

Then, when determining the actual torque, the nut will need to reduce this value by 20 % at the lubricated surfaces and increase by 20 % when treated rudely. But we also get an analytical dependence on the stresses and moments of friction coefficients.

2. For the calculation of stresses and moments, we need to know the average diameter of the thread d_P and the average diameter of the support ring contact nuts d_Γ. Unfortunately, the corresponding coefficients k_P and k_Γ are not constant. In particular, ratios k_0 and k_P depend not only on the thread pitch h and on the diameter of the bolt d. Coefficient k_0, for example, receives the values for a standard metric thread in a range from 0.8 for small, medium and 0.85 to 0.9 for large bolts [5]. Coefficient k_Γ is also slightly dependent on the clearance in the hole δ and the diameter of the bolt due to the standardization of a number of sizes of nuts turnkey is $d_K \approx 1.5d$ [6]. Since the sensitivity of the calculation of these ratios is low, we will rely on the following mean values (for a standard medium standard thread and nut)

$$k_0 = 0.85, \quad k_P = \frac{k_0 + 1}{2} = 0.925, \tag{2}$$

$$k_\Gamma = 1.35. \tag{3}$$

3. We calculate that the strength bolts will hold according to Huber-Mises criterion [7], according to which the equivalent stress is defined by the formula

$$\sigma_{\vartheta} = \sqrt{\frac{(\sigma_1 - \sigma_2)^2 + (\sigma_2 - \sigma_3)^2 + (\sigma_3 - \sigma_1)^2}{2}}, \tag{4}$$

where σ_1, σ_2, σ_3 are the principal stresses. When uniaxial stretching and tightening the bolt (which is our case), the Formula (4) takes the form

$$\sigma_{\scriptscriptstyle 3} = \sqrt{\sigma^2 + 3\tau^2}, \tag{5}$$

where σ—normal (tensile) stress, τ—tangential stress.

4. To ensure the safety margin required under some uncertainty and variation characteristics of materials, assembly parameters and operating conditions, the final test of strength will require

$$\sigma_{\scriptscriptstyle 3}^{\max} \le \frac{\sigma_T}{k_T}, \tag{6}$$

where $\sigma_{\scriptscriptstyle 3}^{\max}$ is the highest possible voltage equivalent to (5), σ_T is the yield strength of the bolt material, k_T is the standard safety factor of yield strength. In engineering, the static load is taken to be [1]:

$$k_T = 1.2 - 2.0.$$

This safety factor should be greater than the accuracy of the calculation which is less and has higher reliability requirements. With dynamic (variable cyclic) loads, the safety factor is calculated to increase by about half [1].

Assume for static loads that the most typical and convenient for theoretical conclusions should be the safety factor of yield strength

$$k_T = \sqrt{2} = 1.414. \tag{7}$$

5. Calculation of bolting is to select the material of the bolt, the permissible minimum diameter of the bolts and nuts of the required torque to ensure reliable operation of bolting under the given longitudinal or transverse loads. Since the diameters of bolts are strictly given, we will solve the inverse equivalent task by the given characteristics of the selected bolt that will determine the maximum permissible load with a specified safety factor (7). If the actual load will be less than the maximum allowable, by the conversion formula we will find the real safety factor, which will be above (7).

Calculation of Pre-stressed State of the Bolt

In all modes of bolting for a joint, a density pre-loaded bolt axial force is created by tightening the nut torque $M_з$. This time depends on the yield strength of the bolt material and is selected depending on the mode of operation so that the initial axial stress in the continuous body of the bolt (in cross-section under the thread) is in the range [1]

$$\sigma_0 = \frac{F_0}{S_0} = (0.4 - 0.6)\sigma_T \tag{8}$$

where F_0 is a thrust prestressed bolt.

Let us find the relationship between M_3 and σ_0 and let us justify the choice of (8). Tightening torque consists of two parts:

$$M_3 = M_\Gamma + M_P, \tag{9}$$

where M_Γ is the friction torque at the end of the nuts and M_P is the friction in the thread (thread time).

Tightening the nut is approximately determined by the formula

$$M_\Gamma = fF_0 \frac{d_\Gamma}{2} = \frac{k_\Gamma}{2} fF_0 d. \tag{10}$$

We need to determine when the thread considers the force of pressure and friction in a pair of metric screw bolt + nut (Fig. 6) to be adequate. There α is an included angle, ψ—elevation angle of the thread. For a metric thread,

$$\alpha = 60°, \quad \psi \approx 2°.$$

We write the equation of the balance of forces on the tangent and the vertical axis.

Fig. 6 Screwing a pair of bolt and nut

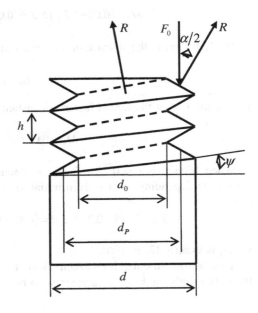

$$F_\tau = fR\cos\psi + R\sin\psi,$$
$$F_0 = R\cos\frac{\alpha}{2}\cos\psi - fR\sin\psi.$$

Here R is a normal reaction in the thread. Except for R, we find the time thread to be

$$M_P = F_\tau \frac{d_P}{2} = \frac{f + tg\psi}{\cos\frac{\alpha}{2} - f\cdot tg\psi} F_0 \frac{d_P}{2}. \tag{11}$$

Note that (11) coincides exactly with the formula obtained in [2].
Then, the tightening torque is

$$M_3 = M_r + M_P = \left(f\frac{k_r}{2} + \frac{f + tg\psi}{\cos\frac{\alpha}{2} - f\cdot tg\psi}\frac{k_P}{2} \right) F_0 d. \tag{12}$$

Expanding (12) in a Taylor series of f in the neighborhood of zero, we obtain

$$M_3 = \left(0.01837 + 1.2098 f + 0.0212 f^2 + 0.00084 f^3 + 3.35\times 10^{-5} f^4 \right) F_0 d.$$

Neglecting nonlinear terms and rounding, we get a simple formula for the torque

$$M_3 = \left(0.02 + 1.2 f \right) F_0 d = \left(0.02 + 1.2 f \right) \sigma_0 S_0 d. \tag{13}$$

Interestingly, in the absence of the thread lifting ($\psi = 0$), series expansion gives

$$M_3 = 1.2 f F_0 d,$$

that is, the rise of the thread brings a constant term

$$M_3 = 0.02 F_0 d.$$

When friction is $f = 0.12 - 0.15$ we obtain a simple approximate formula for the torque depending on the given initial axial stress σ_0:

$$M_3 = \left(0.18 - 0.2 \right)\sigma_0 S_0 d = \left(0.18 - 0.2 \right)\sigma_0 \frac{\pi k_0^2}{4} d^3 \approx 0.1\sigma_0 d^3,$$

which is exactly [2, p. 106].

To determine the maximum equivalent stress in the bolt tightening torque from the initial finds the largest shear stress to be:

$$\tau_0 = \frac{M_P}{W_p}, \tag{14}$$

where W_p is the polar section modulus by thread:

$$W_p = \frac{J_p}{d_0/2} = \frac{\pi d_0^3}{16} = \frac{S_0}{4} d_0, \tag{15}$$

where $J_p = \pi d_0^4/32$—the polar moment of inertia of a thread. Substituting (11) and (15) into (14) we get

$$\tau_0 = \frac{f + tg\psi}{\cos\frac{\alpha}{2} - f \cdot tg\psi} 2\frac{k_P}{k_0}\sigma_0 = \frac{f + 0.0344}{\frac{\sqrt{3}}{2} - 0.0344f} 2.17647 = a\sigma_0, \tag{16}$$

where $a = 0.4662$ when $f = 0.15$.

Thus, the highest equivalent stress in pre-tightening the bolts with the axial stress σ_0 is given by

$$\sigma_э = \sqrt{\sigma_0^2 + 3\tau_0^2} = \sqrt{1 + 3a^2}\,\sigma_0. \tag{17}$$

Expanding the radical expression in (17) in a Taylor series of f in the neighborhood of zero, in view of (16) we obtain a sufficiently high degree of accuracy

$$\sigma_э = \sqrt{1 + 1.3f + 20f^2}\,\sigma_0. \tag{18}$$

With an average level of friction $f = 0.15$ we obtain

$$\sigma_э = 1.3\sigma_0, \tag{19}$$

which is exactly [2, p. 109].

At low friction $f = 0.12$ will have to be

$$\sigma_э = 1.2\sigma_0.$$

If more friction $f = 0.18$ will have to be

$$\sigma_э = 1.4\sigma_0.$$

A safety factor for the yield strength of prestressed bolts is determined by the formula

$$k_T = \frac{\sigma_T}{\sigma_э}. \tag{20}$$

At the maximum delay $\sigma_0 = 0.6\sigma_T$, in the absence of other loads will be at $f = 0.15$, the minimum allowable safety factor

$$k_T = \frac{\sigma_T}{\sigma_3} = \frac{\sigma_T}{1.3 \cdot 0.6\sigma_T} \approx 1.3.$$

Bayonet Connection Mode

Glow mode bolting (Fig. 2) in the absence or low pre-tightening of the bolt (net pin) is the easiest way to calculate the stresses and obtain a maximum external load.

Pure Pin Mode

In the clean pin bolt it works mainly with a shear. The maximum permissible external shear load Q_0 is given in (5) and (6) where the condition

$$\sigma_3 = \sqrt{3}\,\tau \le \frac{\sigma_T}{k_T}, \tag{21}$$

and where the shear stress τ is equal to

$$\tau = \frac{Q_0}{S} = \frac{Q_0}{\pi d^2/4}. \tag{22}$$

Substituting (22) into (21), we obtain the maximum shear external load

$$Q_0 = \frac{1}{\sqrt{3}k_T} S\sigma_T. \tag{23}$$

With $k_T = \sqrt{2}$ we have

$$Q_0 = \frac{1}{\sqrt{6}} S\sigma_T = 0.4 S\sigma_T = 0.1\pi d^2\sigma_T. \tag{24}$$

Note. Calculation of shear bolts by Formulas (21)–(24) is valid provided that the bearing stress does not exceed the equivalent shear stress, i.e. where

$$\sigma = \frac{Q_0}{dh_i} < \sigma_3 = \sqrt{3}\,\tau = \frac{\sqrt{3}Q_0}{\pi d^2/4},$$

where h_i is the minimum thickness of the "shearing" of the details (see Fig. 2). It follows that the condition that must be satisfied is

$$h_i > \frac{\pi d}{4\sqrt{3}} = 0.453d. \tag{25}$$

Inequality (25) is certainly satisfied if the following simple condition holds:

$$h_i \geq 0.5d = r. \tag{26}$$

If the magnitude of shear force Q is less than the maximum allowable (24), then the safety factor for the yield point k_T is recalculated according to (23) according to the formula

$$k_T = \frac{1}{\sqrt{3}} \frac{S\sigma_T}{Q} = \sqrt{2} \frac{Q_0}{Q}. \tag{27}$$

Mode with Pre-tightening

A mode pin bolt connection with pre-tightening of the bolt to the axial stress σ_0 is mainly used for tight and reliable connection parts. They prevent their relative displacement by friction between the connected parts. In this case, the main stress in the bolt will be axially from the pre-tightening and tightening the nuts on the tangent. I.e. stress-strain state of the bolt in this connection will match the state of pre-tightening of the bolts, discussed in Sect. 3. Nevertheless, we find an expression for the equivalent of residual stress at the shearing force Q_1. The exact formula for the equivalent stress in this case is

$$\sigma_{\vartheta} = \sqrt{\sigma_0^2 + 3(\tau_0 + \tau_1)^2}, \tag{28}$$

where τ_0 is shear stress by tightening the nuts (16), τ_1 is the residual shear strain, determined by the formula

$$\tau_1 = \frac{Q_1}{S} = \frac{Q - Q_T}{S} = \frac{Q - f_{\varPi} F_0}{S} = \frac{Q - f_{\varPi}\sigma_0 S_0}{S} \quad \text{with } Q > Q_T, \tag{29}$$

$$\tau_1 = 0 \quad \text{with } Q \leq Q_T,$$

where Q is shear force, Q_T is the force of friction between the connected parts.

When natural restriction $\tau_1 \leq \sigma_0$ is a sufficiently accurate approximation (29) we get that

$$\sigma_3 = \sqrt{1+3a^2}\,\sigma_0 + \frac{3a}{\sqrt{1+3a^2}}\tau_1. \tag{30}$$

The approximate formula for (30) depending on the friction coefficient f is

$$\sigma_3 = \sqrt{1+1.3f+20f^2}\,\sigma_0 + \frac{0.2+8f}{\sqrt{1+1.3f+20f^2}}\tau_1. \tag{31}$$

With an average level of friction $f = 0.15$ we obtain

$$\sigma_3 = 1.3\sigma_0 + 1.1\tau_1. \tag{32}$$

A safety factor of yield strength of a prestressed bolt experiencing residual shear stress, when is $f = 0.15$ determined by the formula

$$k_T = \frac{\sigma_T}{\sigma_3} = \frac{\sigma_T}{1.3\sigma_0 + 1.1\tau_1}. \tag{33}$$

Mode of Longitudinal Confinement

In the longitudinal confinement (native bolting mode, Fig. 3), along with the initial voltage the tightening bolt is subject to an additional axial tensile stress due to the external load N.

To find the highest equivalent stress in the bolt, first we obtain an expression for the full axial force in the bolt from the external load. Due to the flexible element compounds, that effort is not difficult (see e.g. [2]), and is written in the form

$$F = F_0 + \chi N, \tag{34}$$

where F is the total tensile force in the bolt, F_0 is force the axial pre-tensioning bolt, χ is the main factor of the load which is generally defined as

$$\chi = \frac{\lambda_P}{\lambda_B + \lambda_H + \lambda_P}, \tag{35}$$

where λ_B are stretch bolts, λ_H is in compliance of loaded components, λ_P is compliance discharged parts. For compounds of metal components, the main load ratio is in the range [2]

$$\chi = 0.2 - 0.3. \tag{36}$$

Expression (34) is valid until the beginning of the opening at the junction

$$N \leq N^*, \tag{37}$$

where N^* releases the external force, wherein there is an opening of the joint. When $N > N^*$ is the disclosure of the joint, the bolt will take all the external load and total tension force in the bolt as

$$F = N. \tag{38}$$

Liberating the joint force N^* is easily found by equating (34) and (38):

$$N^* = \frac{F_0}{1 - \chi}. \tag{39}$$

Theoretical (under the linear elastic connection elements) dependence of the total force in the bolt from the external load is shown in Fig. 7 (see also [1]).

To prevent disclosure of the joint at the maximum permissible external load N_0, the latter must be slightly smaller than the releasing force N^*. It is convenient to put

$$N_0 = F_0. \tag{40}$$

Then the stock density in accordance with the joint (36) and (39) would be sufficient and be

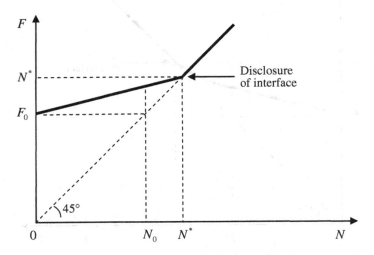

Fig. 7 The theoretical dependence of the total force in the bolt from the external load

$$\frac{N^*}{N_0} = \frac{1}{1-\chi} = 1.25 - 1.43.$$

To clarify the factor, χ construct the dependence of the total force in the bolt from the external load by simulating the package ANSYS works bolting shown in Fig. 8.

Figure 9 shows the results of numerical modeling of a bolted connection in Fig. 8, a very good agreement with the experimental data [3]. The slight difference from the theoretical piecewise linear relationship (Fig. 7) is due to the presence of nonlinear contact relations.

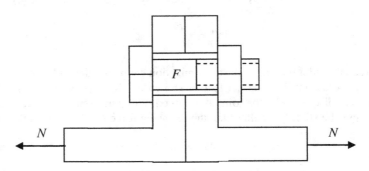

Fig. 8 Simulated bolted connections

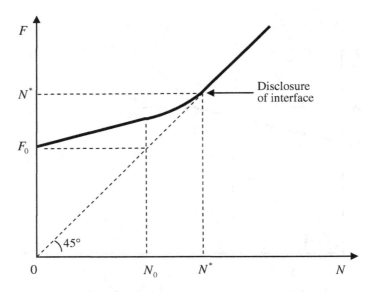

Fig. 9 The experimental dependence of the total force in the bolt from the external load

According to the results of numerical simulation (Fig. 9), we find the average value of the coefficient of refined base load to be

$$\chi = 0.25. \tag{41}$$

With such a load factor of the main joint, stock density will be

$$\frac{N^*}{N_0} = \frac{1}{1 - 0.25} = 1.33, \tag{42}$$

which is sufficient to provide a reliable and tight joint.

Note that the sensitivity of the tension in the bolt to the factor χ is not so great and, therefore, accurate knowledge of χ is not so important.

To determine the maximum external load N_0 and voltage corresponding to the optimum tightening σ_0^{opt}, find the full axial stress in the cross section of the bolt threads. In accordance with (34) we have

$$\sigma = \frac{F}{S_0} = \sigma_0 + \chi\sigma_N, \tag{43}$$

where $\sigma_N = N/S_0$ is load stress, $\chi\sigma_N$ is the stress of the load added to the initial axial stress tightening bolts.

Consider the general case for universality, when the bolt tensile stress is σ_N and shear stress is τ_1. Then the exact formula for the equivalent stress is

$$\sigma_{\ni} = \sqrt{\sigma^2 + 3(\tau_0 + \tau_1)^2} = \sqrt{(\sigma_0 + \chi\sigma_N)^2 + 3(\tau_0 + \tau_1)^2}. \tag{44}$$

At reasonable values $\sigma_N \leq \sigma_0$ and $\tau_1 \leq \sigma_0$ (we recall that the joint tightness is ensured by $\sigma_N \leq \sigma_0$) sufficiently accurate approximation (44) gives

$$\sigma_{\ni} = \sqrt{1 + 3a^2}\,\sigma_0 + \frac{\chi}{\sqrt{1 + 3a^2}}\sigma_N + \frac{3a}{\sqrt{1 + 3a^2}}\tau_1. \tag{45}$$

The approximate formula for (45) depending on the friction coefficient f:

$$\sigma_{\ni} = \sqrt{1 + 1.3f + 20f^2}\,\sigma_0 + \frac{\chi}{\sqrt{1 + 1.3f + 20f^2}}\sigma_N + \frac{0.2 + 8f}{\sqrt{1 + 1.3f + 20f^2}}\tau_1 \tag{46}$$

With an average level of friction $f = 0.15$ we obtain

$$\sigma_{\ni} = 1.3\sigma_0 + 0.8\chi\sigma_N + 1.1\tau_1. \tag{47}$$

Note that for a similar formula [2, p. 115] in the second term there is no factor of 0.8, and the third term is absent.

When $\chi = 0.25$ the Formula (47) takes the simple and convenient for engineering calculations views

$$\sigma_\mathfrak{z} = 1.3\sigma_0 + 0.2\sigma_N + 1.1\tau_1. \tag{48}$$

We find in this case, the optimal tightness of bolts in the absence of shear stress ($\tau_1 = 0$), providing maximum tensile stress ($\sigma_N = \sigma_0$). In accordance with (48)

$$\sigma_\mathfrak{z} = 1.3\sigma_0 + 0.2\sigma_0 = 1.5\sigma_0.$$

If an acceptable safety factor is

$$k_T = \frac{\sigma_T}{\sigma_\mathfrak{z}} = \sqrt{2},$$

then we have

$$\sigma_0{}^{opt} = \frac{\sigma_\mathfrak{z}}{1.5} = \frac{\sigma_T}{1.5\sqrt{2}} = 0.4714\sigma_T \approx 0.5\sigma_T. \tag{49}$$

In the general case of arbitrary stress $\sigma_N \leq \sigma_0$ and $\tau_1 \leq \sigma_0$ with $f = 0.15$ and $\chi = 0.25$ the safety factor is calculated by the formula

$$k_T = \frac{\sigma_T}{1.3\sigma_0 + 0.2\sigma_N + 1.1\tau_1}, \tag{50}$$

where tightening tension σ_0 is selected from the range (8).

Note. In the manufacture of high strength steel bolts, for example, steel 40X with $\sigma_T = 785\,\mathrm{MPa}$ in order to reduce the load on the other structural elements it is recommended to reduce the number of tightening bolts compared to the optimal value (49) to a value

$$\sigma_0 = 0.4\sigma_T. \tag{51}$$

In this case if $f = 0.15$, $\chi = 0.25$ and the external loads are absence then the safety factor has the largest value

$$k_T = \frac{\sigma_T}{1.3 \cdot 0.4\sigma_T} \approx 1.9. \tag{52}$$

The Tight Connection Mode

The tight connection (Fig. 4) experiences only tension bolt pre-tightening. Therefore, to maximize the density of connections and maximum shear force holding Q, the bolt is a tightened maximum torque $M_\mathfrak{z}^{\max}$ until the maximum equivalent stress is reached. According to (18), in this case a low level of friction $f = 0.12$, we have

$$\sigma_{\mathfrak{z}} = 1.2\sigma_0,$$

where the received values are the safety factor for strength $k_T = \sqrt{2}$. In accordance with (20) we get to rounding with

$$\sigma_0^{\max} = \frac{\sigma_T}{1.2k_T} = \frac{\sigma_T}{1.2\sqrt{2}} = 0.59\sigma_T \simeq 0.6\sigma_T. \tag{53}$$

The corresponding maximum torque according to the bolt (18) and (53) will be

$$M_{\mathfrak{z}}^{\max} = 0.093 d^3 \sigma_0^{\max} \approx 0.06 d^3 \sigma_T. \tag{54}$$

Find the largest holding transverse force $Q^{\max}$. It is determined solely by friction between the connected parts. With

$$f_{\mathcal{A}} = 0.18,$$

$$Q^{\max} = Q_T = f_{\mathcal{A}} F_0^{\max} = f_{\mathcal{A}} \sigma_0^{\max} S_0 = 0.18 \cdot 0.6\sigma_T S_0 \approx 0.1\sigma_T S_0 \approx \frac{1}{6} F_0^{\max}.$$

Conclusions

These approximate analytical formulas for calculating bolt connections with strength have simple and compact shapes and are very comfortable in practice, particularly in engineering. The accuracy of the calculations is sufficient. Obtained formulas allow one to quickly define all the characteristics of a given bolting or to pick up the required bolt, providing it has a preset mode loading. The basic formulas are the formulas for finding the required torque (13) and the resulting equivalent stress (46)–(48).

References

1. Birger IA, Shorr BF, Iosilevich GB (1979) Strength analysis of machine parts: handbook. Engineering, Moscow, 702 p (in Russian)
2. Reshetov DN (1989) Machine parts. Engineering, Moscow 496 p
3. Birger IA, Iosilevich GB (1990) Threaded and flanged connections. Engineering, Moscow, 368 p (in Russian)
4. Anurev VI (1979) Reference design-mechanic, 3 Vols, V. 2. Engineering, Moscow, 560 p (in Russian)
5. GOST 24705-2004 (2005) Metric thread. Standartinform, Moscow (in Russian)
6. GOST 10495-80 (1992) Hexagon nuts for flange connections. Standartinform, Moscow (in Russian)
7. Pavlov PA, Parshin LK, Melnikov BE, Sherstnev VA (2003) Strength of materials. Publishing House "Lan", St. Petersburg, 528 p (in Russian)

Modern Methods of Contact Forces Between Wheelset and Rails Determining

Kirill V. Eliseev

Abstract Systems that contain instrumented wheelsets and algorithms of measurements evaluations are used to obtain contact forces between wheels and railway while a car is moving. Existing schemes do not allow us to obtain all contact characteristics, thus a new method of evaluations was developed. It allows evaluation of all contact forces components and contact points coordinates. Numerical experiments were conducted to prove the quality of the method.

Keywords Structural mechanics · Strain measurement · Inverse problem · Simulation · Contact forces · Railway technology

Introduction

Forces between wheels and railway are very important parameters that characterize movement of railway carriages. Contact interaction data can be used to analyze railway quality and new parts performance.

Direct forces measurement is inconvenient, so other values like strains and relative displacements are used with corresponding algorithms of data evaluation. Currently measuring gauges are usually placed on rails or wheelsets. The last variant makes continues analysis for long distances possible, though data treatment is very complicated.

This article contains review of some available measurement schemes, the new one is introduced with some results.

K.V. Eliseev (✉)
Peter the Great Saint-Petersburg Polytechnic University, Saint Petersburg, Russia
e-mail: kir.eliseev@gmail.com

© Springer International Publishing Switzerland 2016
A. Evgrafov (ed.), *Advances in Mechanical Engineering*,
Lecture Notes in Mechanical Engineering, DOI 10.1007/978-3-319-29579-4_6

Review of Available Measurement Schemes

Publications about wheel and rail contact can be divided into two groups. The first one includes analytical and numerical analysis for simplified railway and train models. The most important are publications of Kalker [1], in Russian publications of Romen [2]. Their results can be used to analyze the main phenomena of movement, but real railway features cannot be accounted for.

The aim of the second group of investigations is to analyze contact conditions for a specific train during movement on a specific railway.

According to measurement gauge positions, schemes are divided to be

- installed on rails;
- installed on trains.

The main disadvantage of the first group of measurements is that it can be performed only with a prepared and relatively short piece of the railway. An example of the scheme is installation on both sides of the rails of vertical resistance strain gauges; see Fig. 1a. The sum of strains can be used to get vertical force, the difference being the transverse force.

In [3] piezoelectric gauges are used to monitor railway structures, like bridges, Fig. 1b.

Measurement systems on trains are usually based on wheel strain measurements by means of strain gauges. In this case the following components must be connected to form a measurement system:

- wheelset, a wheel's geometry defines the gauges positions;
- signal measurement equipment with a limited number of channels;
- calibration stand;
- algorithm of contact forces and contact position evaluation.

The simplest schemes are based on relations between one contact force component (usually vertical or axial) with one strain measurement.

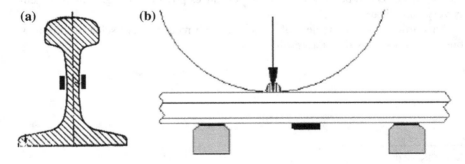

Fig. 1 Railway gauges installation examples

For example, the system in Fig. 2a uses 32 wheel gauges. Some of them are installed inside additional holes. The first half of these gauges is used for vertical force evaluation based on disk compression, the second—for axials based on bending. Evaluations are performed when one of gauge's diameters coincide with a vertical [4].

Patent [5] defines a scheme that includes gauges diametrically connected in half bridges. Gauges are installed on the inner surface of the wheel, Fig. 2b. A synchronization module is responsible for evaluation of a time point when one of the gauge's diameters is vertical, then a corresponding measuring channel is used. Obtained data sets are processed according to a special algorithm.

Another continuous measurement of vertical and axial force's schemes includes two independent bridges. Gauges are located on circles on both sides of a disk with circumferential distance 45° [5]. Any time combination of data (sum or difference) that depends only on one force component is used.

In the study [6], a vertical force component evaluation algorithm is developed. Pairs of strain gauges on both sides of the wheel are installed on one circle with circumferential distance 45°, Fig. 2c. During the calibration procedure a constant

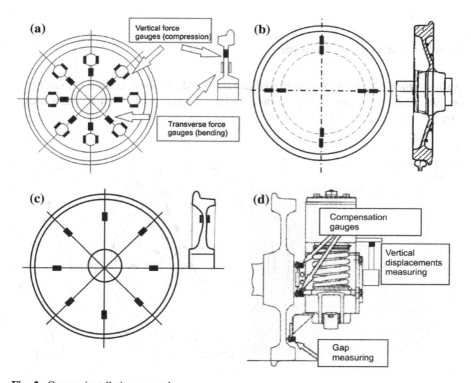

Fig. 2 Gauges installation examples

traveling along the circle force $P0$ is applied. A calibration function is obtained as $A(t) = P0/V0(t)$, where $V0(t)$—the recorded measurement. Then arbitrary force $P(t)$ can be restored using measurement $V(t)$ as $P(t) = A(t)^{-1} \cdot V(t)$.

Operation of the first instrumented wheel sets shows their weakness—a wired signal transmitting from rotating parts to non-rotating ones. Measurements on no rotation bogie parts are suggested in [4]. Axial force is evaluated based on wheel bending according to distance to the disk, see Fig. 2d. Axis-relative vertical translation gives an estimation of vertical force. Additional efforts have been made to reduce errors: compensation gauges are used for gaps, while bearings with reduced gaps to reduce axial translations and additional surface treatment are used.

The evaluations scheme mentioned above in one way or another have the following disadvantages:

- no scheme allows us to evaluate a contact point position and longitudinal force;
- contact point position influence is not accounted for;
- significant wheels and bogies modifications are made that can lead to lower reliability;
- schemes use specific wheels design and can not be used for new ones.

New Measurement Scheme

An instrumented wheelset with curved disks is used. A wheel section is presented on Fig. 3. A radiotelemetering complex with 64 channels is used for data measurement and registration.

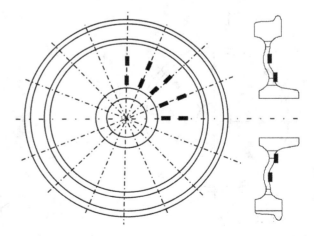

Fig. 3 Gauges on wheel (for 1/4 wheel)

The algorithm is based on the relation between a vector of measured strains increments $\Delta\varepsilon$ and a vector of force components with increments and contact points coordinates ΔR like

$$\Delta\varepsilon = A\Delta R \tag{1}$$

where A is a Jacobi matrix $n \times m$.

Here the number of equations $n = 64$ is higher than the number of unknowns $m = 8$ and a system of equations usually does not have an exact solution. So a "pseudo solution" ΔR is used that minimizes a Euclid norm of error $\|A\Delta R - \Delta\varepsilon\|$,

$$\Delta R = (A^T A)^{-1} A^T \Delta\varepsilon. \tag{2}$$

A finite element model of a wheelset is used to obtain matrix A coefficients. The wheelset is fixed at bearings. Forces and their increments at contact points are applied, strains are calculated on disk surfaces [7].

The following variant of gauges placement was chosen after results were determined of analysis on two circles at the inner surface of every disk with increment 22.5°. Radial strains are used. This placement scheme accounts for:

- most sensitive to forces variations areas;
- ability to restore strains distribution along circle;
- ability to restore wheelset rotation angle.

Numerical experiments to restore contact forces and coordinates in static were performed. Different combinations of forces and coordinates were used with satisfactory results. A calibration scheme that includes a calibration stand was introduced that allows one to make precise matrix A coefficients [7, 8].

Accounting for Wheel Inertia Effects

Development of contact loads evaluation algorithms based on strains measurement usually account for only static loading behavior of a wheelset. Calibration usually is available only for static loading. Stands that can be used for dynamic modeling are much more expensive. Procedures testing for dynamic loading can be performed only using train movement models.

It is very likely that during train movement there will exist dynamic loads due to bogie movement on rails and defects of rails and wheels [6].

If a contact forces spectrum contains harmonics with frequencies, close to wheel eigenfrequencies, corresponding strains harmonics will have higher values than in static loading. Then forces evaluated using static calibration will have overestimated values.

For a considered wheelset the first eigenfrequency is 152 Hz, the third is 223 Hz. Special attention must be paid to modes 9–12 with frequencies 958 and 1416 Hz— significant radial strains exist where gauges are installed, Fig. 4.

Numerical experiments were performed with a finite element model of one wheel. Harmonic loads were applied at contact points with low frequencies (20 and 50 Hz), near resonance (151 Hz) and post resonance (175 Hz).

Results significantly depend on damping coefficients α and β, that defines a damping matrix in the equation of motion

$$M\ddot{U} + B\dot{U} + CU = F \qquad (3)$$

Fig. 4 Wheel eigenmodes, axial displacements

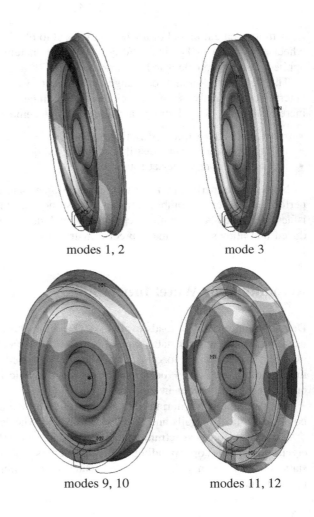

modes 1, 2   mode 3

modes 9, 10   modes 11, 12

where

M mass matrix,
C stiffness matrix,
B damping matrix, $B = \alpha M + \beta C$,
U nodal displacements vector,
F nodal forces vector.

Parameters α and β can be used to set a prescribed damping ratio for force frequencies range $[\omega_1, \omega_2]$. Parameter α governs damping for lower frequencies, β for higher.

Some numerical harmonic calculations were performed: different harmonic contact forces were defined and some combinations of parameters α and β were used. Figure 5 presents results for an applied vertical force with amplitude of 100KN.

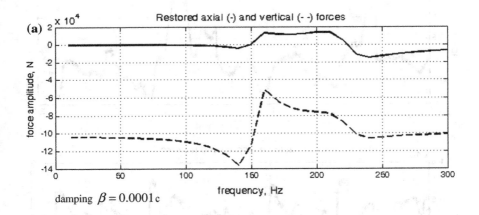

damping $\beta = 0.0001\,c$

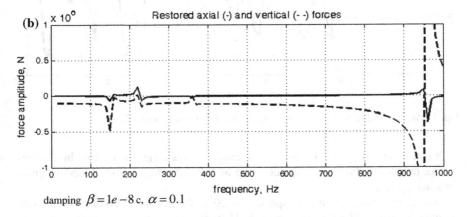

damping $\beta = 1e-8\,c,\ \alpha = 0.1$

Fig. 5 Results for vertical and axial harmonic force restored. **a** Damping $\beta = 0.0001c$. **b** Damping $\beta = 1e - 8c,\ \alpha = 0.1$

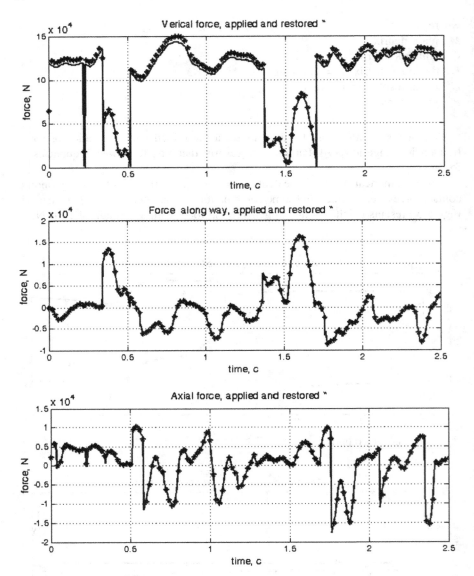

Fig. 6 Applied and restored (*asterisk* every 20th point shown) contact forces

It can be stated that wheel material damping plays a significant role in a forces restore procedure. For low frequencies up to 100–120 Hz, a static procedure gives satisfactory results.

During train movement with speed 80–120 km/h, periodical contact forces with frequencies up to 110 Hz are expected. They are due to wheels rotation, sleepers positions, long rails defects (300–450 mm).

In monograph [9] spectral densities of forces and accelerations are presented for train ER200. Significant harmonics have frequencies 1–3 Hz, some local maximums for 3–16 Hz.

Thus it can be expected that significant force harmonics will lie inside an allowable range and will be restored.

A series of numerical experiments was conducted to restore forces that were evaluated for models of bogies 18-9810 and 18-9855 [10]. Figure 6 presents some examples of force versus time dependencies, Fig. 7—force amplitudes spectra in case of movement along a straight railway.

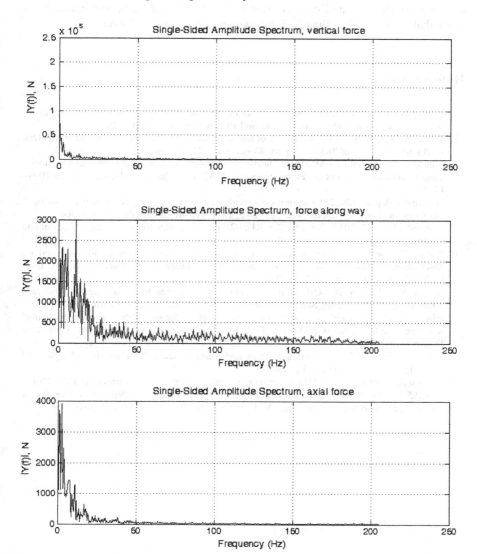

Fig. 7 Forces amplitudes spectra

Significant harmonics have frequencies up to 50–100 Hz. The scheme allows us to restore forces with desirable accuracy.

Conclusion

The article presents a brief review of known schemes of wheel rail contact forces restoration. New algorithm of forces evaluation is presented that allow us to obtain all force components and contact coordinates. Results of numerical experiments show that errors are less than 5 % for dynamically changing forces.

References

1. Kalker JJ (1973) Simplified theory of rolling contact. Delft progress report. Series C: mechanical and aeronautical engineering and shipbuilding, vol 1, pp 1–10
2. Romen SU (1969) Study of side effects of rolling stock on the way with the use of computers. Works of VNIIZT, vol 385. Transport, Moscow, pp 71–94 (rus)
3. Sekula K, Kolakowski P (2013) Identification of dynamic loads generated by trains in motion using piezoelectric sensors. In: Proceedings of ISMA2010 including USD201, 2013, pp 1099–1118
4. Matsumoto A [etc.] (2012) Continuous observation of wheel/rail contact forces in curved track and theoretical considerations. Veh Syst Dyn: Int J Veh Mech Mobility 50:349–364
5. Krasnov OG [etc.] (2010) Pat. 2441206 RU, 02.11.2010 Device for measuring of vertical and lateral forces between wheel and rail
6. Ronasi H, Nielsen J (2013) Inverse identification of wheel–rail contact forces based on observation of wheel disc strains: an evaluation of three numerical algorithms. Veh Syst Dyn: Int J Veh Mech Mobility 51(1):74–90
7. Eliseev KV, Ispolov IG, Orlova AM (2013) Contact forces between wheelset and rails determining. St. Petersburg State Polytechnical Univ J 4-1(183):262–270 (rus)
8. Eliseev K, Migrov A,Orlova A (2012) Design of test-rig for the calibration of instrumented wheelsets (rus). Transport problems 2012. Silesian University of Technology Faculty of Transport, pp 474–479 (rus)
9. Lvov AA [etc.] (1978) Interaction of track and rolling stock at high speeds and high axial loads. Transport, Moscow, 133 p (rus)
10. Saidova AV, Orlova A (2013) Development of mathematical models of cars carts 18-9810 and 18-9855 for the study of wheel wear. Bulletin of Dnepropetrovsk National University of Railway Transport named after acad. V. Lazaryana, vol 2(44), pp 118–123 (rus)

A Novel Design of an Electrical Transmission Line Inspection Machine

Mohammad Reza Bahrami

Abstract This article is aimed at modeling of transmission line inspection robot in order to improve the mechanical mechanism and achieving dynamical stability to navigate through overhead electrical transmission lines while passing obstacles. A new mechanical mechanism design is developed that allows robot navigate more stable than current commercial inspection systems. The behavior of investigation machine while passing obstacle for four critical modes as follows has been studied: (1) Moving on a cable with maximum slope of 30° (2) Passing over aircraft warning lights (3) Passing over aircraft clamps and dampers (4) Passing from the strain insulators with/without twist in the horizon plane.

Keywords Investigation machine · Electrical transmission line

Introduction

Nowadays high-voltage transmission lines play important roles in human life. Electric power is transmitted from generators to cities and industrial centers through transmission lines. Consequently, if any damage and disruption happened to these lines, human life and industries could face a number of problems. In order to prevent further damages in the mentioned areas, inspection and maintenance of high-voltage transmission lines are necessary and important. On the other hand, inspection of high-voltage transmission lines performed by human forces would be faced with danger, but by improvement of technology, robots have been used as reliable machines instead of human forces in potentially dangerous environments such as hot lines. Considering the fact that repair works are often complex to be accomplished by a robot on hot lines, power companies have mainly used robots for automating inspection tasks.

M.R. Bahrami (✉)
Peter the Great Saint-Petersburg Polytechnic University, Saint Petersburg, Russia
e-mail: mr.bahrami@inbox.ru

© Springer International Publishing Switzerland 2016   67
A. Evgrafov (ed.), *Advances in Mechanical Engineering*,
Lecture Notes in Mechanical Engineering, DOI 10.1007/978-3-319-29579-4_7

Robotic inspection means the usage of autonomous or remotely controlled machines that combine imaging, sensing, and other technologies to diagnose the condition and situation of transmission line components. The idea is to decrease or eliminate human exposure to potentially dangerous environments while collecting the data required for inspection of transmission lines in order to decrease the cost of repair. Transmission lines are faced with a variety of factors, which cause different problems and limit lifetime of the lines, such as corrosion and wind induced vibrations.

In the past two decades, some efforts have been made to make fully autonomous and intelligent robots for inspection electrical transmission lines. The simplest power lines have one conductor per phase hung on insulator strings, which can be either suspension or strain insulators. Besides insulators, there are other obstacles on the conductors, such as dampers, aircraft warning lights, and clamps. These robots are able to pass the line equipment and the tower with obstacle avoidance capability by using necessary sensors for hot line inspection. The other challenge is that in windy climates, and even during navigation, sometimes the captured images of the line became blurry, then in most of the proposed designs the image processing which is used for navigation of a robot may be faced with problems.

The robot travels suspended from the conductor and has to cross obstacles along the power line that requires complex robotic mechanisms including conductor grasping systems and robot driving mechanisms. Moreover, an obstacle detection and recognition system, robot control system, communication, inspection platform equipped with necessary sensors and measurement devices, power supply and electromagnetic shielding have to be considered in robot mechanism design and construction. The robot's mechanical mechanism as the main part of the robot design may significantly affect other issues in the whole design process, such as energy consumption and inspection data quality.

The first prototype of inspection robots was proposed to inspect telephone-lines by Aoshima et al. [1]. As an initial prototype, it suffered several limitations such as complexity of its control system and its low speed of movement. Sawada et al. [2] designed a robot in order to inspect the optical fiber lines. The latter robots were capable of moving on the lines with 30° slope and pass the mast tips, however, it still lacked the speed of motion and stability. Higichi et al. [3] suggested a more advanced robot for inspection of power lines. They tried to solve the stability problem while passing over the obstacles, although they still had problems when passing over clamps and other complicated hurdles. Tsujimura et al. [4] designed a wired—suspended mobile robot for inspection of telecommunication cables. Their robot was able to pass different kinds of obstacles in a snake-like motion on the wires using linkage mechanism. Although their robot was ideal because of its constant speed, it was not able to pass mast tips with changing of line direction. One of the most advanced power transmission line inspection robots was LineScout developed by Montambault and Pouliot in 2007 [5]. This robot was able to pass different kinds of obstacles on the phase lines with high-speed motion. However, in spite of high performance of the robot in field tests for moving on straight transmission lines, it still required more improvements in order to pass the mast tips with change of line direction. A variety of other electrical line inspection robots with different purposes are introduced and described in [6–9] for interested readers [10].

According to the mentioned issues, a novel and simple design for a kind of power line inspection machine is needed. In order to accomplish this in the next sections design of the robot, included environment design assumption has been reviewed.

Description of a Novel Electrical Transmission Line Investigation Machine

In order to design a mechanical mechanism for an inspection robot, firstly it is necessary to have some knowledge about the environment where the robot is desired to work.

The Environment Design

As one knows, power lines are also complicated environments and navigation for robots is difficult. In the simplest power lines one can see only one conductor per phase. Hung on insulator strings, these insulators can be either suspension or strain. Besides insulators, other types of obstacles on the conductors also exist, such as clamps, dampers, aircraft warning lights etc. (Fig. 1).

In this article, assumption of environment design parameters are shown in Table 1.

Basic Mechanical Concepts

The basic mechanical schematic of the robot can be seen in Fig. 2. A Diagnostic Machine has been considered by having around three independent frames (Fig. 2): the wheel frame (blue parts), which includes two couple motorized traction wheels,

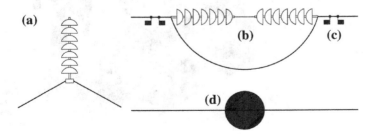

Fig. 1 Different obstacles on electrical line conductors: **a** suspension insulator, **b** strain insulator, **c** damper, **d** aircraft warning sphere

Table 1 Design parameters

Line components	Value
Conductor diameter	15–50 mm
Maximum obstacle length	0.76 m
Number of conductors	1
Maximum slope in span	30°

the arm frame (brown parts), with one arms and one grippers, and the center frame (grey box), which links together the first two frames (called "main body") and allows them to slide and rotate. In the proposed robot, active and passive mechanisms enable it to move over various obstacles on wires, such as dampers, clamps, warning balls etc.

The main body has an important role in a system of a mobile platform since the movement of the two other frames is generated by sliding them in opposite directions. Also it should be mentioned that about 40 % of the platform's weight is related to this part. Therefore, the main body has to be stiff for expected platform behavior as flexible enough to adapt to particular situations, like when the robot is faced with a change in direction of the conductor at a suspension clamp.

The passive mechanisms also include a set of spring dampers installed in each joint of robot arms.

The behavior of a robot while passing an obstacle at five critical modes as follows has been considered:

Fig. 2 Robot schematic

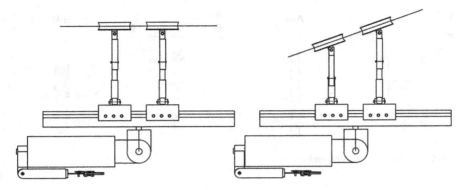

Fig. 3 Moving on a cable with maximum slope of 30°

1. **Moving on a cable with maximum slope of 30°**

   Figure 3 shows mounting of the robot on a conductor with the slope. As shown in Fig. 3 by rotating wheels and using passive mechanisms, the main body remains horizontal, which leads to an improved stability and mobility at any cable slope. Also in this situation, in order to prevent the robot from sliding on the conductor, a spring mechanism has been used between couple wheels that can produce enough interaction force.

2. **Passing over aircraft warning lights**

   The robot can pass the warning ball in the eight-stage process (Fig. 4). As a warning ball is reached, the arm frame is used. In this case the arm and gripper with the help of one of a couple of wheels can temporarily support the robot while the other wheels are transferring to the other side of the obstacle (Fig. 4a–d). To accomplish this, the wheels themselves are needed to flip down under the obstacle. Then the arm frame transfers to the other side of the robot, and in its last

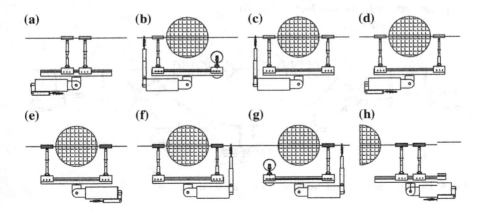

Fig. 4 Passing from aircraft warning balls

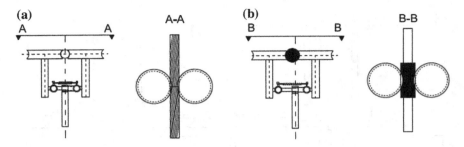

Fig. 5 Passing over clamps and dampers

stages, with the help of one of the coupled wheels, it can temporarily support the robot while the other wheels are transferring to the other side of the obstacle as shown in Fig. 4e–h.

3. **Passing over clamps and dampers**

   The robot can pass the clamps and dampers by using passive mechanisms. As shown in Fig. 5 as one of these kinds of obstacles is reached, wheels are separated from each other. Meanwhile the interaction force between the wheels and conductor produced by passive systems (spring between wheels links) is enough in order to prevent from disconnecting wheels and conductor.

4. **Passing from the strain insulators with/without twist in the horizon plane**

   As a strain insulator is reached, the arm frame is used. In this case the arm and gripper can temporarily support the robot while the wheels are transferring to the plane vertical to the conductor with the help of rotary joints located on the main body. When the wheels are grabbing the conductor, then the arm frame will return to its stationary position to the robot, and the robot will continue the movement as movement on the conductor that has a slope. When the robot comes up from the conductor and reaches the other strain insulator, the arm and gripper can temporarily support the robot while the wheels are transferring to the plane vertical to the conductor (Fig. 6).

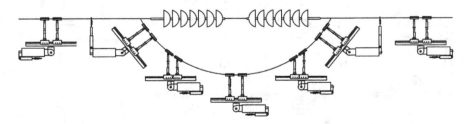

Fig. 6 Passing from the strain insulators twist in the horizon plane

Conclusion

A new model of transmission line inspection robot is proposed for improving the mechanical mechanism and achieving dynamical stability to navigate through overhead electrical transmission lines while passing obstacles. A new mechanical mechanism design is being developed that allows a robot to navigate with more stability than current commercial inspection systems. In the proposed robot, active and passive mechanisms will enable it to move over various obstacles on wires, such as dampers, clamps, warning balls etc. The behavior of the investigation machine while passing an obstacle for four critical modes as follows has been studied: (1) Moving on a cable with maximum slope of 30° (2) Passing over aircraft warning lights (3) Passing over aircraft clamps and dampers (4) Passing from strain insulators with/without twist in the horizon plane.

Acknowledgments The author would like to sincerely thank professor V.V. Eliseev who has supported this research.

References

1. Aoshima S, Tsujimura T, Yabuta T (1989) A wire mobile robot with multi-unit structure. In: Proceedings of the IEEE/RSJ international workshop on intelligent robots and systems, pp 414–421, Sept 1989
2. Sawada J, Kusumoto K, Munakata T, Maikawa Y, Ishikawa Y (1991) A mobile robot for inspection of power transmission lines. IEEE Trans Power Delivery 6(1):309–315
3. Higuchi M, Maeda Y, Tsutani S, Hagihara S (1991) Development of a mobile inspection robot for power transmission lines. J Rob Soc Jpn 9(4):57–63, 1991
4. Tsujimura T, Morimitsu T (1997) Dynamics of mobile legs suspended from wire. Robot Auton Syst 20(1):85–98
5. Montambault S, Pouliot N (2007) Design and validation of a mobile robot for power line inspection and maintenance. In: Proceedings of the 6th international conference on field and service robotics (FSR), July 2007, pp 1–10
6. Toussaint K, Pouliot N, Montambault S (2009) Transmission line maintenance robots capable of crossing obstacles: state-of-the-art review and challenges ahead. J Field Robot 26(5): 477–499
7. Pouliot N, Montambault S (2006) LineScout technology: development of an inspection robot capable transmission and distribution construction, operation and live-line maintenance, 2006, ESMO 2006. IEEE11th international conference
8. Zhou F, Xiao H, Wu A (2006) Control strategy and implementation of an inspection robot for 110 kV power transmission lines, Dalian, China, 2006. 6th World congress on intelligent control
9. Debenest P, Guarnieri M, Takita K, Fukushima EF, Shigeo H, Tamura K, Kimura A, Kubokawa H, Iwama N (2008) Robot for inspection of transmission lines 2008 IEEE international conference on robotics and automation, Pasadena
10. Mostashfi A, Fakhari A, Badri MA (2013) A novel design of inspection robot for high-voltage power lines. World Acad Sci Eng Technol 76

One Stable Scheme of Centrifugal Forces Dynamic Balance

Vladimir I. Karazin, Denis P. Kozlikin, Alexander A. Sukhanov
and Igor O. Khlebosolov

Abstract The paper deals with dynamically balanced centrifugal forces of inertia in rotary shakers. It is considered as a stable scheme of an automatic unloading rotor vibrator mounted on the platform of a centrifugal test stand.

Keywords Centrifugal test stand · Vibration · Stable · Balancing the forces of inertia

Statement of the Problem and the Problems of Implementation

Test the strength of the various devices in the mechanical vibration carried out on special vibrating tables, split time exposure to the test product into two groups: short-acting and long-acting. The former, in particular, are vibroimpact mechanical displays, detailed in [1–4]. The disadvantages of such stands are the lack of the possibility of imposing a large vibration constant acceleration. Such an opportunity is feasible in the stands of the second group. In particular, the rotary vibrating tables are functioning and the problems arise in the present study.

A rotational or centrifugal shaker is represented in Fig. 1. The spindle 1 motor rotates the D with an angular velocity Ω providing a predetermined linear centripetal acceleration A_0. At the end of plate 2 at a distance R from the axis of

V.I. Karazin (✉) · D.P. Kozlikin · A.A. Sukhanov · I.O. Khlebosolov
Peter the Great Saint-Petersburg Polytechnic University, Saint Petersburg, Russia
e-mail: visv05@mail.ru

D.P. Kozlikin
e-mail: kozlikindenis@gmail.com

A.A. Sukhanov
e-mail: Alexeevich@post.ru

I.O. Khlebosolov
e-mail: khlebosolov@mail.ru

© Springer International Publishing Switzerland 2016
A. Evgrafov (ed.), *Advances in Mechanical Engineering*,
Lecture Notes in Mechanical Engineering, DOI 10.1007/978-3-319-29579-4_8

rotation mounted electromechanical vibrator 3, the radial vibration table vibrator creates a mass m_0 on which the product taken by the test is mass m_1. The elastic member of the vibrator has a small rigidity c, so that its natural frequency is below the operating frequency range of the vibrator.

Featured centrifugal shakers act under sufficiently stringent conditions for their operation as well as an extremely broad range of frequencies and accelerations is required. Thus, for example, a considered shaker should provide linear acceleration in the range 10–$150g$ ($g = 9.81$ m/s^2—free fall acceleration), which is superimposed on the radial vibration with a frequency of 5 to 2000 Hz at accelerations in a vacuum table $100g$ with a maximum amplitude of up to 20 mm. WE get the maximum electromagnetic force of the vibrator 50 kN, the mass of products 0–50 kg, and the radius of the location of the unit $R = 2.5$ m. The natural frequency of the vibrator is approximately equal to 3 Hz.

The consequence of these broad requirements for a shaker is complete loss of its original classical scheme shown in Fig. 1. To see this we find the static displacement x_0 of an empty platform of the vibrator by the centrifugal force of inertia at the maximum linear acceleration $A_0 = 150g$. First, we find the rigidity of the elastic element of the vibrator c on its own frequency k.

$$c = k^2 m_0 = (3 \cdot 2\pi)^2 50 = 17765 \, \text{N/m}.$$

Then

$$x_0 = \frac{F_0}{c} = \frac{m_0 A_0}{c} = \frac{50 \cdot 150 \cdot 9.81}{17765} = 4.14 \, \text{m},$$

which is absolutely unacceptable! Note that the table of the vibrator with the product will move even further!

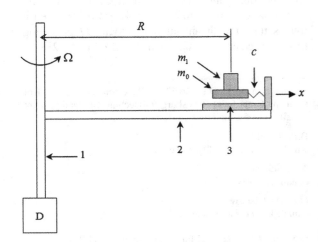

Fig. 1 Centrifugal shaker

Traditional Circuit Static Unloading Centrifugal Forces of Inertia

To reduce the static deflection of the table there are some simple oscillator circuit dynamic unloading of centrifugal forces of inertia. Some of them are discussed in [5–12]. The basic idea is to attach to a table a vibrator rod 4 (Fig. 2) corresponding to the mass m of the counterweight, which creates a compensating force to unload $-F_0$.

At equality of arms of the vibrator and a counterweight (distances R from the respective centers of mass from the axis of rotation), the mass of the counterweight should be equal to

$$m = m_0 + m_1. \tag{1}$$

However, the problem is not limited to the selection of the exact mass of the counterweight (1). More relevant is the problem of a possible loss of stability of the rotating masses. We write the equation of the relative motion of the mass along the radius of the rotating platform 2, subject to the balance of the shoulders and the masses (1) of the rotor vibration of the vibrator and without the installed product:

$$2m_0\ddot{x} = -cx + m_0\Omega^2(R+x) - m_0\Omega^2(R-x), \tag{2}$$

where x is the radial deviation of the position of the vibrator table of dynamic equilibrium. Cutting the same magnitude and opposite in direction of the force of inertia, we obtain the equation

$$2m_0\ddot{x} + m_0(k^2 - 2\Omega^2)x = 0, \tag{3}$$

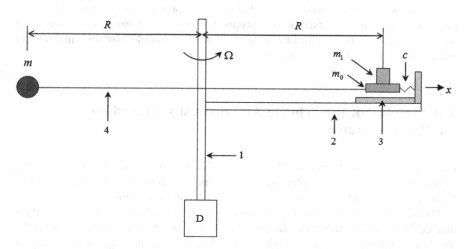

Fig. 2 Unloaded centrifugal shaker

which agrees qualitatively with [13, p. 41]. Stability of equilibrium depends on the sign of the generalized dynamic stiffness $C = m_0(k^2 - 2\Omega^2)$. When $C < 0$ buckling occurs and can provoke an emergency, which is absolutely unacceptable!

The critical rotation speed of the stand conditions $C = 0$. It is

$$\Omega_{kp} = \frac{k}{\sqrt{2}} = \frac{3}{\sqrt{2}} = 2.12\,\text{Hz}. \tag{4}$$

We now define the desired rotation speed of the stand for maximum linear acceleration $A_0 = 150g$:

$$\Omega_0 = \sqrt{\frac{A_0}{R}} = \sqrt{\frac{150 \cdot 9.81}{2.5}} = 24.26\,\text{rad/s} = 3.86\,\text{Hz} > \Omega_{kp} \tag{5}$$

Thus, even with an empty table, a large linear acceleration is guaranteed by the loss of stability. When installed on a table vibrator, the test product's natural frequency of the rotor of the vibrator is reduced and the critical spindle speed stand also will decrease, thereby reducing the operating range of linear acceleration.

Another possible scheme for unloading a dynamic shaker is installed instead of, or in addition to a permanent counterweight that is a lighter air spring vibrator rotor [6]. However, inevitably and undesirably, the natural frequency of the vibrator and the rotor increases and, in addition, will increase the force required for vibration of the electromagnet, which the vibrator cannot give. These nonlinear elastic characteristics of the air spring exacerbates the problem of stability and makes the prospect of using a pneumatic unloading very elusive. Note that the use of these metal springs worsens due to their having movable weight distribution and a wide range of natural frequencies.

These disadvantages of the known schemes of dynamic compensation of centrifugal forces of inertia shakers require the development of new schemes. Dynamic Balance shakers are fundamentally resistant not only to the initial conditions, but also to changes in the parameters of the system (for example, variable mass of the test article or the speed of rotation).

The Stability of the Scheme Automatically Unloading the Rotor Vibrator

Figure 3 shows one of the possible schemes of automatic unloading of the rotor by centrifugal force of the vibrator inertia providing stability of a stationary equilibrium in a rather wide range of parameters and initial conditions.

Unlike previous schemes Fig. 2 shows that the discharge pattern is complemented by a rigid rectangular triangle ABC 5, and is pivotally connected with the rod 4 at a right angle C. This weight is placed in the counterweight m at the top of

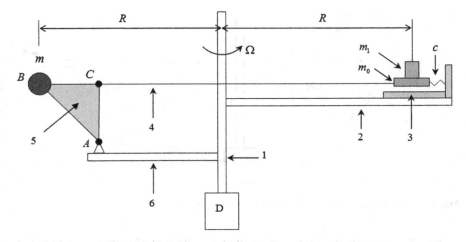

Fig. 3 Sustained self-balanced centrifugal shaker

the triangle B at the same distance R from the axis rotation. A lower apex of the triangle A is pivotally fastened to another platform 6, and rotates together with the spindle 1. This BC cathetus of the triangle is a continuation of the rectilinear rod 4 and the cathetus AC parallel to the axis of rotation in an initial state.

Qualitative resistance in the above diagram of unloading is explained as follows. Suppose that for some reason a table with a vibrator mounted product will move, for example, to the right to a certain amount x. Then approximately the same magnitude and point C is displaced by triangle 5. When the latter rotates about the hinge A it moves clockwise to an appropriate angle. But then, in addition to the centrifugal inertial force applied to the counterweight m (slightly reduced), it thus restores the torque applied to the triangle 5 counterclockwise. This point will return to the starting position of the triangle, and with it the vibrator table with the test product.

If you change the weight m_1 of the product, discharge stability of the scheme is maintained and, only slightly changes the position of the equilibrium: Triangle 5 turns so that the total effect of inertial forces and moments acts on the counterweight as compensated inertial force which is acting on the rotor of the vibrator.

This scheme allows automatic unloading and, if necessary, holds the vibrator rotor on one and the same distance from the axis of rotation of products with different masses. To do this properly we adjust the mass of the counterweight or move the hinge A to platform 6.

To accurately determine the stability conditions of the proposed scheme unloading, turn to Fig. 4. Here m is plenty of table vibration mounted with the test product and the mass of the counterweight placed at the same distance R from the axis of rotation as the vibrator table x is table radial displacement of the vibrator from the equilibrium position, $\Delta\alpha$ are these angles of rotation of the triangle ABC, the corresponding radial displacement x.

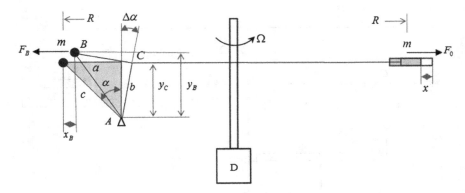

Fig. 4 The deviation from the equilibrium position

We find the generalized dynamic stiffness C deflection table on the value of the vibrator x.

The resulting restoring force is then equal to

$$F = -Cx = -cx + F_0 - F_C, \qquad (6)$$

where F_0 is the centrifugal force of inertia acting on the vibrator table:

$$F_0 = m\Omega^2 (R + x), \qquad (7)$$

where F_C is the centrifugal force of inertia acting on the counterweight, but reduced to the point C of ABC. The latter is converted from the inertial force at point B

$$F_B = m\Omega^2 (R - x_B) \qquad (8)$$

by equating the moments of forces F_B and F_C relative to point A:

$$F_B y_B = F_C y_C. \qquad (9)$$

Necessary variables are found in the assumption that the radial displacement of the vibrator table x:

$$\Delta\alpha = \frac{x}{b}, x_B = c\Delta\alpha \cdot \cos\alpha = x, y_B = b + c\Delta\alpha \cdot \sin\alpha = b + \frac{a}{b}x, y_C = b. \qquad (10)$$

Substituting (10) into (6)–(9) and neglecting infinitely small second-order term x^2, we obtain

$$F = -\left[c + m\Omega^2 \left(\frac{Ra}{b^2} - 2\right)\right]x, \qquad (11)$$

where we find a generalized dynamic stiffness, which is the sustainability of the proposed scheme of dynamic unloading of centrifugal force that must be positive:

$$C = c + m\Omega^2 \left(\frac{Ra}{b^2} - 2 \right) > 0. \tag{12}$$

For (12), only if the following simple and easily feasible conditions for stability,

$$Ra > 2b^2. \tag{13}$$

For example, when $a = b$ (the triangle ABC is isosceles) enough to counter-balance the radius of the location would be more than two legs of the triangle ABC:

$$R > 2a.$$

There is a very important advantage of the proposed scheme of Dynamic Balance inertial forces over traditional schemes. A sufficient condition for the stability of (13) is exclusively geometric. This means that the stability is not affected by the rigidity of the vibrator, audio weight products or the rotation speed of the stand! Moreover, in contrast to the linearized model, based on which the stability conditions were obtained, the actual scheme is non-linear, which ensures absolute stability and full security in any proportion of variable parameters by a hard peg hinge A to the rotating platform 6. However, as already noted, the change of mass products would lead to only a slight deviation of the equilibrium position, which, however, may optionally be offset by adjusting the mass of the counterweight or corresponding displacement of the hinge A.

Finally, if the condition of mass balance (1) with equal arms of the vibrator and counterweight equilibrium position of the vibrator table remain at any spindle speed stand, this will be significant when tested at various speeds.

Simulation of Circuits of Automatic Unloading of the Rotor from the Inertial Forces of the Vibrator

To confirm the efficiency of the proposed scheme of automatic unloading of the rotary test stand by the centrifugal forces of inertia, we simulate operation of the mechanism shown in Fig. 3, the application of Solid Works Motion of Solid Works. This application allows one to use a numerical integration method to analyze movement of the coupled system of bodies.

Reproduction in Solid Works three-dimensional model as an unloading device is shown in Fig. 5.

This model consists of a fixed support 1, with respect to which the platform 2 is rotated at a predetermined angular velocity Ω. At the end of platform 2, at a distance R from the axis of rotation of the installed mass of 3 (m), it imitates the rotor with a

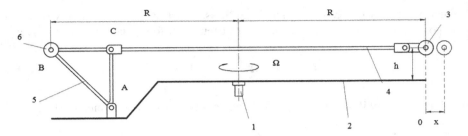

Fig. 5 Model shaker in solid works

vibrator mounted on a test object. The weight 3 is connected with the rod 4, which in turn connects to a rigid rectangular triangle 5 (ABC). Items 3, 4 and 5 are connected between a rotational kinematic pairs. The top of the triangle A is pivotally secured to the platform 2. The apex B at a distance R from the axis of rotation of the counterweight 6. The mounted model provides that translational movement of the mass 3 always occurs strictly parallel to the plane of plate 2 (h = const). Weight 3 under the action of a harmonic force of the vibrator can reciprocally move along the radius of the stand. The origin of this movement (0) coincides with the position of dynamic equilibrium.

Initial data for the following simulation. Consider a Weight table vibrator mounted with a product and the mass of the counterweight $m = 50$ kg, radius placing table vibrator and counterweight $R = 2.5$ m, the rotation speed of the stand $\Omega = 20$ rad/s, which corresponds to a linear constant acceleration $A_0 = 102g$.

When exactly as masses m and shoulders R, centrifugal forces are balanced and the system will be in a state of dynamic equilibrium, which is confirmed by simulation. The corresponding schedule does not mean it is uninformative.

To determine the stability of the equilibrium state of the vibrator, reject the table on the right by the amount $x = 10$ mm. The simulation results (the coordinates of the vibrator table from time to time) are shown in Fig. 6.

The resulting dependence of the vibrator table shows that the system established stable free oscillations with a frequency corresponding to the generalized dynamic stiffness (12).

To eliminate the continuous oscillations in dissipation it is appropriate to make, for example in the form of friction, kinematic pairs of nodes (as in the model and the real stand). The result account of friction is presented in Fig. 7.

After we have seen the stability of dynamic equilibrium of the system, you can "turn on" vibrator, creating a reciprocating table mounted with a product. Enclose in Solid Works Motion to table the vibrator to the amplitude of the harmonic force of 100 N and a frequency of 8 Hz. The result of such a complex spatial modeling is shown in Fig. 8. It is clearly seen that after the decay transients are forced vibration table vibrators with a predetermined frequency.

Now consider the case of a mass unbalance vibrator. Let the mass of the vibrator to the mass of the product be per 1 kg counterweight. The simulation results of

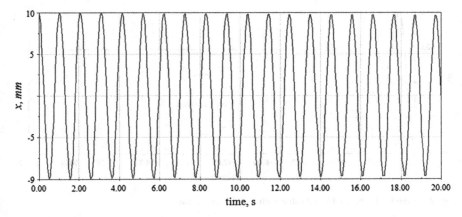

Fig. 6 Fluctuations vibrator table when his initial rejection

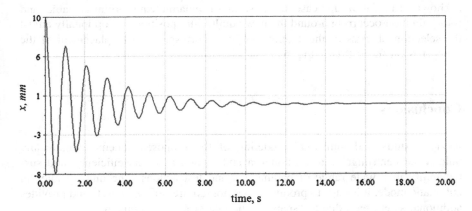

Fig. 7 Damped oscillations of the vibrator table with friction

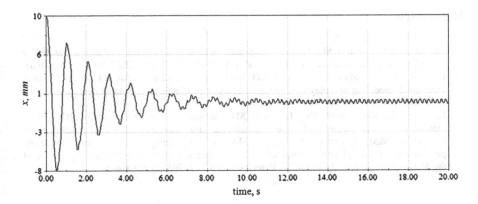

Fig. 8 Forced vibration table vibrator

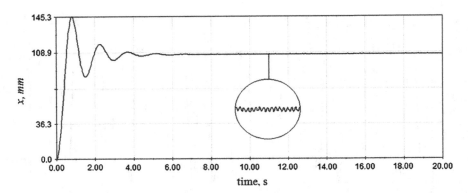

Fig. 9 Forced vibration table vibrator with unbalanced mass

forced oscillations of the vibrator table with Dissipation at an initial rejection $x = 0$ is shown in Fig. 9. In this case, the position of dynamic equilibrium is stable, and rotor vibration occurring around the new equilibrium position can optionally adjust the selection of mass of the counterweight or corresponding displacement of the support A on frame 2 (see Fig. 5).

Conclusions

Analytic study and numerical modeling of the proposed scheme dynamically unload the centrifugal forces of inertia and demonstrates its efficiency and sustainability in a wide range of parameters, and test actions. The unit operates in automatic mode, it is easy to produce, does not require serious presets and provides additional safety tests due to "snap" to the platform at the point A.

References

1. Yarovitsin VS, Litvinov SD, Karazin VI, Sukhanov AA, Khlebosolov IO (2009) The device for testing of products on vibro-impact load. PATENT # 2348021, 27.02.2009 (in Russian)
2. Karazin VI, Kolesnikov SV, Litvinov SD, Sukhanov AA, Khlebosolov IO (2013) Features vibroshock modeling and playback effects. Theory of mechanisms and machines. Periodic Sci Methodical J 2(22):55–64, V. 11 (in Russian)
3. Karazin VI, Kolesnikov SV, Litvinov SD, Sukhanov AA, Khlebosolov IO (2013) Optimization parameters of the broadband mechanical vibroshock stand. In: Radkevich MM, Evgrafov AN (eds) Modern engineering. Science and education: proceedings of the 3rd international scientific and practical conference. Publishing House of the Polytechnic University Press, St. Petersburg, pp 752–765 (in Russian)

4. Vibration technique. Directory in 6 Vols. /Ed. Tip: V.N.Chelomei (chairman), Engineering, Moscow, 1981, V. 4. Vibrating machines and processes/Ed. Lavendel EE, 1981, 510 p (in Russian)
5. Karazin VI, Kozlikin DP, Sloushch AV, Khlebosolov IO (2007) Dynamic model of vibratory stand. Theory of mechanisms and machines. Periodic Sci Methodical J 1(9):38–44, V. 5 (in Russian)
6. Karazin VI, Kozlikin DP, Khlebosolov IO (2007) On balancing the inertial forces in vibro sentrifugal. Theory of mechanisms and machines. Periodic Sci Methodical J 2(10):63–71, V. 5 (in Russian)
7. Vibration technique. Directory in 6 Vols. /Ed. Tip: V.N.Chelomei (chairman), Moscow: Engineering, 1981, V. 5. Measuring and testing /Ed. M.D.Genkin, 1981, 496 p (in Russian)
8. Anuryev VI (1979) Directory design-mechanic, 3 Vols, V. 3. Engineering, Moscow, 560 p (in Russian)
9. Klyueva VV (ed) (1982) Test equipment handbook, 2 books. Engineering, Moscow, Book 1, 528 p (in Russian)
10. Bashta TM (1972) Hydraulic and pneumatic automation: textbook for universities. Engineering, Moscow, 320 p (in Russian)
11. Evgrafov AN, Karazin VI, Smirnov GA (1999) Roller stands for playback of motion parameters. Scientific and technical statements of SPbSTU, vol 3(17). SPbSTU Press, St. Petersburg, pp 89–94 (in Russian)
12. Evgrafov AN, Karazin VI, Khlebosolov IO (2003) Playing motion parameters on rotary stands. Theory Mech Mach 1(1):92–96 (in Russian)
13. Panovko YaG (1980) Introduction to mechanical vibrations. Science, Moscow, 272 p (in Russian)

New Effective Data Structure for Multidimensional Optimization Orthogonal Packing Problems

Vladislav A. Chekanin and Alexander V. Chekanin

Abstract A new multilevel linked data structure designed for fast construction of packing schemes during solving of orthogonal packing problems is considered in this paper. The computational experiments that were carried out demonstrate the high time efficiency of the proposed data structure compared to an ordered simple linked list that requires ordering of all its elements after placing a new object into a container.

Keywords Orthogonal packing problem · Data structure · Multilevel linked data structure · Resource allocation · Optimization

Introduction

Orthogonal packing problems are related to NP-completed problems of combinatorial optimization. Solution of a large number of practical problems of resource allocation in industry and engineering leads to solving orthogonal packing problems [1]. In particular, this problem takes a place in solving of such important problems as optimal filling up of containers, logistics planning, transportation loading, traffic and calendar planning, waste minimization in cutting and many others [2–6].

Consider the statement of the D-dimensional orthogonal packing problem. Here are a set of N orthogonal containers (D-dimensional parallelepipeds) with the dimensions $\{W_j^1, W_j^2, \ldots, W_j^D\}, j \in \{1, \ldots, N\}$ and a set of n orthogonal objects (D-dimensional parallelepipeds) with the dimensions $\{w_i^1, w_i^2, \ldots, w_i^D\}, i \in \{1, \ldots, n\}$. We denote the position of an object i in a container j by $(x_{ij}^1; x_{ij}^2; \ldots; x_{ij}^D)$. In this

This work was carried out with the financial support of the Ministry of Education and Science of Russian Federation.

V.A. Chekanin (✉) · A.V. Chekanin
Moscow State University of Technology "STANKIN", Moscow, Russia
e-mail: vladchekanin@rambler.ru

A.V. Chekanin
e-mail: avchekanin@rambler.ru

© Springer International Publishing Switzerland 2016
A. Evgrafov (ed.), *Advances in Mechanical Engineering*,
Lecture Notes in Mechanical Engineering, DOI 10.1007/978-3-319-29579-4_9

87

problem it is necessary to place all objects into the minimal number of containers on performing of the following conditions of the correct placement [2, 7]:

1. all edges of the packed orthogonal objects are parallel to the edges of the orthogonal containers;
2. all packed objects do not overlap each others, i.e.,

$$\forall j \in \{1, \ldots, N\}, \forall d \in \{1, \ldots, D\}, \forall i, k \in \{1, \ldots, n\}, i \neq k$$

$$\left(x_{ij}^d \geq x_{kj}^d + w_k^d\right) \vee \left(x_{kj}^d \geq x_{ij}^d + w_i^d\right);$$

3. all packed objects are within the bounds of the containers, i.e.,

$$\forall j \in \{1, \ldots, N\}, \forall d \in \{1, \ldots, D\}, \forall i \in \{1, \ldots, n\} \ \left(x_{ij}^d \geq 0\right) \wedge \left(x_{ij}^d + w_i^d \leq W_j^d\right).$$

To solve all the orthogonal packing problems it is common to use various heuristic and metaheuristic optimization algorithms [6, 8–10].

Multilevel Linked Data Structure for Orthogonal Packing Problems

Papers [4, 11] show an efficiency of the proposed model of potential containers intended to describe packing schemes that are generated during solving the multidimensional orthogonal packing problems. In this model any free space of a container is described by a set of potential containers in the form of orthogonal objects with the largest dimensions, placed at some points of the container, with no overlap by them and with all packed into the container objects and the edges of the container. Each potential container k is described with a vector $\{p_k^1; p_k^2; \ldots; p_k^D\}$ containing its dimensions as well as with a vector $\{x_k^1; x_k^2; \ldots; x_k^D\}$ containing coordinates of one of its points which is nearest to the origin of a container which it contains.

A statement of the multidimensional orthogonal packing problem includes specifying a direction of container load as the priority selection list $L_P = \{P_1; P_2; \ldots; P_D\}$ of the coordinate axes, where $P_d \in [1; D] \ \forall d \in [1; D]$. To achieve placement of all objects into a container in a predetermined load direction all the potential containers are chosen according to the priority selection list L_P.

For managing of sets of potential containers we offer a new data structure named as a multilevel linked data structure. The basis of the developed data structure is the idea of presenting a set of coordinates of potential containers as a set of recursively, embedded each to the others, ordered linear queues.

A set K of potential containers located in points $\{x_k^1; x_k^2; \ldots; x_k^D\}, k \in K$ in the multilevel linked data structure is represented as a D-level, recursively embedded

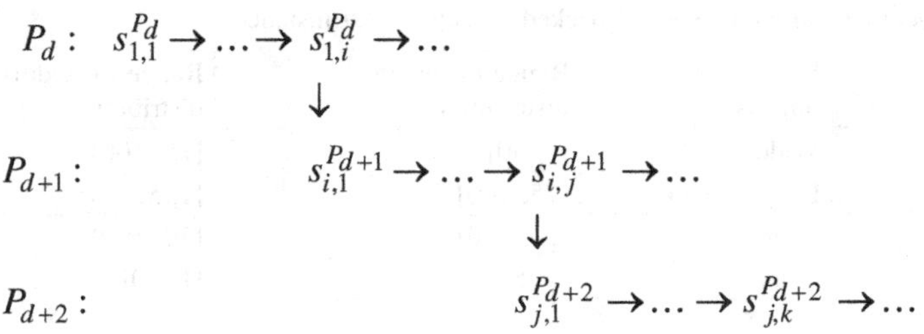

$$P_d: \quad s_{1,1}^{P_d} \to \ldots \to s_{1,i}^{P_d} \to \ldots$$
$$\downarrow$$
$$P_{d+1}: \qquad\qquad s_{i,1}^{P_d+1} \to \ldots \to s_{i,j}^{P_d+1} \to \ldots$$
$$\downarrow$$
$$P_{d+2}: \qquad\qquad\qquad s_{j,1}^{P_d+2} \to \ldots \to s_{j,k}^{P_d+2} \to \ldots$$

Fig. 1 Multilevel linked data structure

Table 1 Original coordinates of three-dimensional potential containers

Number of a point (k)	1	2	3	4	5	6	7	8	9	10
Coordinate x_k^1	0	2	2	2	4	0	4	0	4	2
Coordinate x_k^2	1	3	7	9	1	2	1	2	1	7
Coordinate x_k^3	0	1	5	6	2	1	3	3	1	2

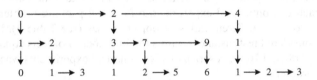

Fig. 2 Multilevel linked data structure generated for a load direction $L_P = \{1; 2; 3\}$

each to the other linear queues, are ordered by increase of their items as it shown on Fig. 1. Each item j of a queue i on a level P_d contains a coordinate $s_{i,j}^{P_d} = x_k^{P_d}$ of such a potential container k that within each queue the inequality $s_{i,j}^{P_d} < s_{i,j+1}^{P_d} \forall P_d \in L_P$ must be satisfied. Sorting of all items in the multilevel linked data structure is performed automatically by inserting a new element into a position list, which satisfies this inequality.

For example we consider a three-dimensional orthogonal container in the form of a parallelepiped that contains a set K of potential containers described by vectors $X_k = \{x_k^1; x_k^2; x_k^3\}, k \in K$ as given in Table 1. The multilevel linked data structure for a load direction $L_P = \{1; 2; 3\}$ for this case is shown on Fig. 2.

Computational Experiment

The effectiveness of the proposed multilevel linked data structure is investigated on the standard two-dimensional orthogonal bin packing problems taken from the OR-library (http://people.brunel.ac.uk/~mastjjb/jeb/info.html). OR-library is a test

Table 2 Geometrical parameters of packed objects in test instances

Class of objects	Feature of objects	Range of length distribution	Range of width distribution
Class 1	Wide	[1, 50]	[75, 100]
Class 2	Long	[75, 100]	[1, 50]
Class 3	Large	[50, 100]	[50, 100]
Class 4	Small	[1, 50]	[1, 50]

library for a variety of operations research problems. These two-dimensional standard test instances 2DBPP (2D Bin Packing Problem) were proposed by Fekete and Schepers [12].

Computational experiments were carried out using a developed application software Packer [9] on a personal computer (CPU—AMD 1.79 GHz, RAM—1.12 GB). This software is specially created to solve a variety of cutting and packing problems requiring optimization of orthogonal resources.

In all instances containers are the same and have dimensions 100×100. All objects in considered packing problems are grouped into four classes given in Table 2. In each computational experiment, orthogonal packing problems of three types were solved with different ratios of object classes (see Table 3). For each of this type we considered test instances with a total number of objects equal to 40, 50, 100, 150, 250, 500 and 1000. A series of 100 numerical experiments was performed for each test problem.

The efficiency of the new data structure is estimated by relative time efficiency T that is calculated by formula

$$T = \frac{t_l}{t_m} \times 100\,\%,$$

where t_l and t_m—time was spent on placing all the objects when using the simple linear linked list and at using the multilevel linked data structure, respectively.

Figure 3 gives a diagram, based on the averaged test results obtained for all types of considered two-dimensional orthogonal packing problems (ngcutfs 1-3). This diagram shows that the proposed multilevel linked data structure provides faster

Table 3 Types of test packing problems

Types of problems	Ratios of object classes, %			
	Class 1	Class 2	Class 3	Class 4
Type 1 (ngcutfs 1)	20	20	20	40
Type 2 (ngcutfs 2)	15	15	15	55
Type 3 (ngcutfs 3)	10	10	10	70

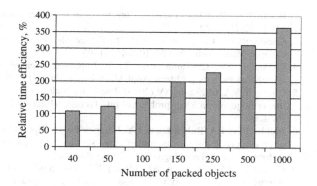

Fig. 3 Relative time efficiency of multilevel linked data structure for 2DBPP

placement of objects compared to the simple linear linked list which uses for sorting the algorithm Quicksort—one of the fastest known sort algorithm [13].

Results of carried out computation experiments on solving three-dimensional orthogonal packing problems that demonstrate the efficiency of application of the multilevel linked data structure are given in paper [14]. The results are the same as obtained here showing that the time efficiency of a multilevel linked data structure increases with the number of objects. This data structure is the most effective to solve packing problems with a large number of objects belonging to several object types of slightly different size.

Conclusion

A new effective data structure applicable to any dimensional orthogonal packing problems has been proposed. The developed multilevel linked data structure is based on the idea of representation of a set of coordinates of potential containers as recursively embedded, each to the others, ordered linear linked lists. The depth of this data structure is equal to the dimension of the considered orthogonal packing problem.

The computational experiments carried out on standard instances of two-dimensional orthogonal packing problems demonstrated the high efficiency of the proposed data structure compared with the simple ordered linear linked list. Multilevel linked data structures increase the speed of construction of two- and three-dimensional packing schemes more than twice in practice.

References

1. Wascher G, Haubner H, Schumann H (2007) An improved typology of cutting and packing problems. EJOR 183(3):1109–1130

2. Bortfeldt A, Wascher G (2013) Constraints in container loading—A state-of-the-art review. EJOR 229(1):1–20
3. Chekanin AV, Chekanin VA (2013) Efficient algorithms for orthogonal packing problems. Comput Math Math Phys 53(10):1457–1465
4. Chekanin AV, Chekanin VA (2013) Improved packing representation model for the orthogonal packing problem. Appl Mech Mater 390:591–595
5. Chekanin VA, Chekanin AV (2013) Based on a multimethod technology algorithm for solving the orthogonal packing problems. Inf Technol (Informacionnye Tehnologii) 7:17–21 (in Russian)
6. Mukhacheva EA, Bukharbaeva LY, Filippov DV, Karipov UA (2008) Optimization problems of transport logistics: operating bin stowage for cargo transportation. Inf Technol (Informacionnye Tehnologii) 7:17–22 (in Russian)
7. Chekanin VA, Chekanin AV (2012) Effective models of representations of orthogonal resources in solving the packing problem. Inf Control Syst (Informatsionno-Upravlyayushchiye systemy) 5:29–32 (in Russian)
8. Chekanin VA, Chekanin AV (2012) Researching of genetic methods to optimize the allocation of rectangular resources In: Modern engineering: science and education: proceedings of 2nd International scientific conference. Izd-vo Politekhn. un-ta, SPb, pp 798–804 (in Russian)
9. Chekanin VA, Chekanin AV (2012) Optimization of the solution of the orthogonal packing problem. App Inf (Prikladnaya informatika) 4:55–62 (in Russian)
10. Crainic TG, Perboli G, Tadei R (2008) Extreme point-based heuristics for three-dimensional bin packing. INFORMS J Comput 20(3):368–384
11. Chekanin VA, Chekanin AV (2013) A model of managing free spaces of containers for the orthogonal packing problem In: Modern engineering: science and education: proceedings of 3th international scientific. Izd-vo Politekhn. un-ta, SPb, pp 1060–1065 (in Russian)
12. Fekete SP, Schepers J (1998) New classes of lower bounds for bin packing problems. Integer programming and computational optimization (IPCO 98). Lect Notes Comput Sci 1412:257–270
13. Weiss MA (2014) Data Structures and algorithm analysis in C++. Boston, Pearson Education, 656 p
14. Chekanin VA, Chekanin AV (2014) Improved data structure for the orthogonal packing problem. Adv Mater Res 945–949:3143–3146

One-Dimensional Models in Turbine Blades Dynamics

Vladimir V. Eliseev, Artem A. Moskalets and Evgenii A. Oborin

Abstract Turbine blades are considered as straight naturally twisted rods. Two models are discussed: Bernoulli–Euler beam and Cosserat rod. Linear theories with small displacements, rotations and loads are used. The equations of dynamics taking into account bending, twisting, axial and shear deformations and cross links between them are derived. The stiffness coefficients in elasticity relations are defined. In the case of harmonic oscillations, we have for amplitudes the ordinary differential equations solved by means of computer mathematics (Mathcad). As a result the normal modes, the natural frequencies and also the amplitudes of forced oscillations are obtained. For the Bernoulli–Euler beam the Lagrange–Ritz–Kantorovich variational approach with approximations of deflections is proposed. The unknown coefficients of approximation depending on time are found by the numerical integration of the Lagrange system of equations. Proposed methods are applied to calculation of the real turbine blade.

Keywords Turbine blades · Naturally twisted rods · Linear rod theory · Elastic modulus · Natural frequencies · Normal modes · Variational approach · Forced oscillations · Computer mathematics

Introduction

The analysis of turbine blades stress-strain state continues to attract the attention of mechanical engineers [1–7], but the focuses in that area are shifted. Initially turbine blade calculations of the beams [8–10] predominated, but then finite element

V.V. Eliseev (✉) · A.A. Moskalets · E.A. Oborin
Peter the Great Saint-Petersburg Polytechnic University, Saint-Petersburg, Russia
e-mail: yeliseyev@inbox.ru

A.A. Moskalets
e-mail: artem.moskalec@gmail.com

E.A. Oborin
e-mail: oborin1@yandex.ru

© Springer International Publishing Switzerland 2016
A. Evgrafov (ed.), *Advances in Mechanical Engineering*,
Lecture Notes in Mechanical Engineering, DOI 10.1007/978-3-319-29579-4_10

analysis of three-dimensional models began to displace them [11]. However, the scopes of one-dimensional models have not been exhausted. By using them, problems that were unassailable in a three-dimensional formulation [12, 13] can be solved. Considering rod models is justified if the peculiar properties of the three-dimensional stress-strain state have not appeared [14, 15].

The goal of this work is the creation of engineering methods of turbine blade calculation with two one-dimensional models [16, 17]: the Bernoulli–Euler beam (with bending in two planes) and the Cosserat rod. For the beam, the variational method of Lagrange [18] with its combination of computer mathematics [19] is used, allowing us to solve also some difficult nonlinear oscillation problems [17].

The full one-dimensional model of a turbine blade as a twisted rod [20–22], taking into account bending, twisting and axial deformations as interconnected, is more complex. The cross link between bending and twisting arises from the difference between the center of stiffness and the center of mass. The cross link between extension and twisting arises from natural twisting. The analysis of such a difficult model became possible due to progress in elastic rod mechanics [18, 23–25] and computer mathematics.

Blade as Beam with Bending in Two Planes

We involve the triple of Cartesian axes directing the z-axis through the cross section centers of mass (we assume that the centers of mass lie on the same line). The displacement of a twisted rod always has two components u_x, u_y, which are the functions of coordinate z and time t. The kinetic energy is the integral along the blade length:

$$T = 1/2 \int_0^l \rho \left(\dot{u}_x^2 + \dot{u}_y^2 \right) dz \tag{1}$$

where ρ is the mass distributed per unit length.

We define potential energy in the usual way for the beam theory assumption, that the axial displacement at the bending is $u_z = -xu'_x - yu'_y$ (here prime is the derivative with respect to z):

$$\Pi = E/2 \int_0^l dz \int_F u_z'^2 dF = E/2 \int_0^l \left(J_x u_x''^2 + 2J_{xy} u_x'' u_y'' + J_y u_y''^2 \right) dz \tag{2}$$

(here E is Young's modulus). We introduce the moments of inertia as integrals $J_x = \int x^2 dF$, $J_{xy} = \int xy dF$, $J_y = \int y^2 dF$ over the cross section F.

From the model with distributed parameters we proceed to the discrete one by means of displacements approximation

$$u_x(z,t) = \sum_{k=1}^{n} U_{xk}(t)\varphi_k(z) = U_x(t)^T \varphi(z), \ u_y(z,t) = U_y(t)^T \varphi(z) \qquad (3)$$

(with matrix notation). The functions $U_{x,y}(t)$ characteristic to the Kantorovich method [18] play the role of generalized coordinates and are subject to the definition by the Lagrange equation system. The coordinate functions $\varphi(z)$ are given according to the boundary conditions (fixed at $z = 0$). We accept

$$\varphi_i(z) = z^{1+i}, \quad i = 1, \ldots, n. \qquad (4)$$

Substituting formulas (3) and (4) into the definitions (1) and (2), we obtain the kinetic energy and the potential energy of the blade discrete model:

$$T = 1/2\left(\dot{U}_x^T m \dot{U}_x + \dot{U}_y^T m \dot{U}_y\right), \quad m = \rho \int_0^l \varphi \varphi^T dx;$$

$$\Pi = 1/2\left(U_x^T C_x U_x + 2U_x^T C_{xy} U_y + U_y^T C_y U_y\right), \qquad (5)$$

$$C_x = E \int_0^l J_x \varphi'' \varphi''^T dz, \ C_{xy} = E \int_0^l J_{xy} \varphi'' \varphi''^T dz, \ C_y = E \int_0^l J_y \varphi'' \varphi''^T dz$$

with the stiffness matrix and the inertia matrix.

The generalized forces for the Lagrange equations are found by the virtual work:

$$\int_0^l \left(f_x \delta u_x + f_y \delta u_y\right) dz = Q_x^T \delta U_x + Q_y^T \delta U_y, \ Q_x = \int_0^l f_x \varphi dz. \qquad (6)$$

The column Q_y is founded similarly. Here f_x, f_y are the components of the distributed per unit length load.

Lagrange Equations and Solving of Them

Reducing the notation (1–3) by involving block columns and block matrixes

$$U = \begin{pmatrix} U_y \\ U_z \end{pmatrix}, \ M = \begin{pmatrix} m & 0 \\ 0 & m \end{pmatrix}, \ C = \begin{pmatrix} C_x & C_{xy} \\ C_{xy} & C_y \end{pmatrix}, \ Q = \begin{pmatrix} Q_x \\ Q_y \end{pmatrix}, \qquad (7)$$

we write the Lagrange equations

$$\left(\frac{\partial K}{\partial \dot{U}}\right)^{\cdot} - \frac{\partial K}{\partial U} = -\frac{\partial \Pi}{\partial U} + Q \Rightarrow M\ddot{U} + CU = Q(t). \qquad (8)$$

In the case of free oscillations ($Q = 0$), the main oscillations are considered with the normal modes Φ and the natural frequencies λ:

$$U(t) = \Phi \sin \lambda t; \quad (C - \lambda^2 M)\Phi = 0. \qquad (9)$$

Here is the generalized eigenvalue problem solved by the Mathcad built-in functions [19].

The normal modes Φ_i have the well-known properties of orthogonality. Normalizing the modes we obtain $\Phi_i^T M \Phi_k = \delta_{ik}$, $\Phi_i^T C \Phi_k = \lambda_i^2 \delta_{ik}$. Expanding the displacement according to the modes, we arrive to the main coordinates V_i:

$$U(t) = \sum_{i=1}^{2n} V_i(t)\Phi_i = \Gamma V, \ \Gamma = (\Phi_1 \dots \Phi_n) \qquad (10)$$

The basis matrix Γ is composed of the mode columns. With the main coordinates we write

$$K = \frac{1}{2}\sum \dot{V}_i^2, \ \Pi = \frac{1}{2}\sum \lambda_i^2 V_i^2, \ \underline{\ddot{V}_i + \lambda_i^2 V_i = P_i(t)}, \ P = \Gamma^T Q. \qquad (11)$$

The underlined equation is solved with Duhamel's integral

$$V_i = V_i(0)\cos \lambda_i t + \dot{V}_i(0)\lambda_i^{-1}\sin \lambda_i t + \lambda_i^{-1}\int_0^t P_i(\tau)\sin \lambda_i(t - \tau)d\tau. \qquad (12)$$

The formulas (3), (10) and (12) give the solution of the forced oscillation problem if the frequencies and modes are known.

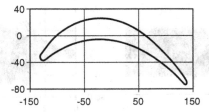

Fig. 1 The blade cross section in mm

The calculation is simplified if computer mathematics (Mathcad) is utilized. The set of frequencies and the basis matrix Γ are found by the built-in functions genvals and genvecs.

The calculations of the specific blade produced by one of the companies of Russia are done. The length of the blade $l = 0.78$ m, the cross section is presented in Fig. 1. The cross section profile is bounded by the graphs of two functions $y_l(x)$, $y_t(x)$ defined by points with regression (the built-in functions regress and interp):

The approximation of displacements (3) with the number of terms $n = 5$ is accepted. Then the number of degrees of freedom, of the frequencies and of the modes is $2n = 10$. In Fig. 2 the results for the first six modes are presented:

Further we discuss an example of calculation of transient forced oscillations. The distributed per unit length loads are given by

$$f_{x,y}(z,t) = f_{x,y}(z)\cos(\omega(t)t), \quad \omega(t) = \lambda_0 + \dot{\lambda}t \tag{13}$$

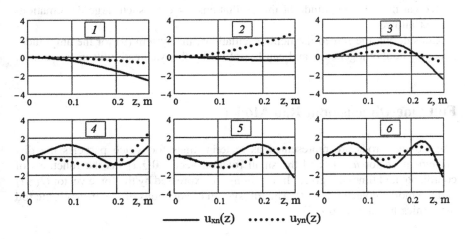

Fig. 2 The normal modes

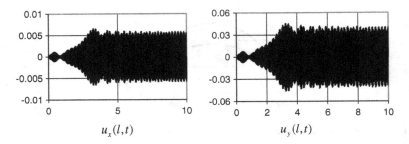

$u_x(l,t)$ $u_y(l,t)$

Fig. 3 The transient oscillations

where $f_{x,y}(z)$ are the presented blade aerodynamics results, and $\lambda_0, \dot{\lambda}$ are given constants. We have an abruptly applied load with a pass through the resonance, and λ_0 is less than one of natural frequencies. The calculation according to the above discussed method does not cause difficulties, and the integrals (12) are computed by means of Mathcad.

Note, that due to computer mathematics the preliminary calculation of the frequencies and the modes ceases to be compulsory—the system of ordinary differential Eq. (8) (ODE) can be integrated right off. We represent this system

$$\dot{U} = v, \quad \dot{v} = M^{-1}(Q - CU) \tag{14}$$

and appeal to the built-in function Radau. This function is a special numerical method for solving the stiff systems of ODE [19]. The conventional Runge–Kutta methods turn out to be ineffective for the system (14). In Fig. 3 results for the pass through the first resonance with $f_x(z) = 1000 = -f_y(z)$, $\lambda_0 = 470$, $\dot{\lambda} = 2$ are presented (in m). The amplitude of the oscillations ≈ 4 cm—such large deformations are not allowed.

Considering the forced harmonic oscillations, the system (9) for the amplitudes becomes nonhomogeneous and can be solved by the Mathcad built-in function.

Full One-Dimensional Rod Model

The Cosserat rods are represented as material lines for which particles the displacement vector $u(z,t)$ and the small rotation vector $\theta(z,t)$ (as the functions of coordinate and time) are given. Interactions are expressed by the force vector $Q(z,t)$ and the moment vector $M(z,t)$. The full system of equations of the linear elastic rod mechanics has the form [18]:

$$Q' + q = \rho \ddot{u}, \quad M' + k \times Q + m = G \cdot \ddot{\theta},$$
$$\theta' = A \cdot M + C \cdot Q, \quad u' = \theta \times k + B \cdot Q + M \cdot C. \tag{15}$$

The first two equations express the laws of balance of momentum and angular momentum. In them ρ is the distributed per unit length mass, G is the inertia tensor, k is the tangent unit vector (z-axes). The third and the fourth equations are the elasticity relations connecting the deformation vectors θ' and $\gamma = u' - \theta \times k$ with the force factors (the stress resultants). Note that the deformation vector θ' defines curving and twisting of the rod, and the vector γ defines extension (compression) and transversal shear. In elasticity relations we have three tensors of compliance: A is the compliance in bending and twisting, B is the compliance in tension and shear and C is the tensor of cross links.

The distributed per unit length force q and moment m are the applied external loads in (15). They are assumed to be small in the considered linear theory as like as u, θ, Q, M. Then G, A, B, C correspond the initial (unstrained) state and do not depend on time.

The tensors of compliance A, B, C are founded by means of the theory of elasticity from the solution of Saint-Venant problem or (which is more difficult) from the asymptotic analysis of the three-dimensional problem with small thickness [19]. The calculation of the compliance tensors is a separate topic beyond the scope of this work.

The model with the Eq. (15) is called full because it describes bending, twisting and extension of the rod as interconnected deformations. Defined by tensor C cross links arise from shifting of the mass center and from natural twisting. The consequence of this links is the simultaneous bending, twisting and tension of the blade practically under any load.

Further the full one-dimensional blade model, the method of calculation using computer mathematics and the estimation of the link's influence on the stress-strain state are presented.

The principal directions of the inertia tensor G are rotated with regard to initial ones (with $z = 0$) by an angle $\varphi(z)$. The twist $\Omega = \varphi'(z)$ is involved. For the straight rod from the generalized Saint-Venant problem, it is established [22]:

$$A = (EI)^{-1} + A_z kk, \quad B = (\mu FK)^{-1} + B_z kk, \quad C = A_z k\eta + C_z kk \tag{16}$$

Here $J = \int xx dF$, $J = \mathrm{tr} J$, $I = -k \times J \times k$ are the geometrical moments of inertia (x is the position vector in the cross section); we see the well-known expressions of compliance in bending $(EI)^{-1}$ and extension (compression) $B_z = (EF)^{-1}$.

In relations (16) the geometrical twisting stiffness C_Φ, the position vector of the center of bending η and the coefficient of cross link between extension and twisting

C_z are included. The determination of them is associated with the solving of problem of twisting [18, 26]:

$$\Delta\Phi = -2, \Phi|_{\partial F} = 0; \quad \Delta W = 0, \partial_n W = \mathbf{n} \times \mathbf{x} \cdot \mathbf{k};$$
$$C_\Phi = 2\int \Phi dF, \quad A_z = (\mu C_\Phi)^{-1},$$
$$C_z = \Omega(\mu F)^{-1}(1 - JC_\Phi^{-1}),$$
$$\boldsymbol{\eta} = J^{-1} \cdot \int \mathbf{x} W dF. \tag{17}$$

We can determine the warping function W without solving the Neumann boundary value problem (17) if the Cauchy–Riemann equations for conjugate harmonic functions are used:

$$\nabla W = \nabla\left(\Phi + \frac{x^2}{2}\right) \times \mathbf{k} \Rightarrow W = \int\left[(\partial_y\Phi + y)dx - (\partial_x\Phi + x)dy\right]. \tag{18}$$

Also in Eq. (16) K is the tensor of shear factors. Its determination is related to the solution of two more boundary value problems for the vector fields in the cross sections [18, 22]. In this article K is the identity tensor.

To determine the stress function Φ and the warping function W, we use the variational method [26]. As above we introduce two functions $y_t(x)$, $y_l(x)$ whose graphs bound the cross section top and bottom. We find the approximate solution in the form $\Phi = \alpha(y - y_t(x))(y - y_l(x)) = \alpha\Phi_0$. The unknown varied factor α is founded by the minimization of the functional

$$J = \int_F \left[|\nabla\Phi|^2 - 4\Phi\right]dF = \int_{x_0}^{x_1} dx \int_{y_l(x)}^{y_t(x)} \left(\alpha^2|\nabla\Phi_0|^2 - 4\alpha\Phi_0\right)dy \rightarrow \min. \tag{19}$$

Then the geometric twisting stiffness is $C_\Phi = 2\alpha\int_{x_0}^{x_1} dx \int_{y_l(x)}^{y_t(x)} \Phi_0 dy$. Further we define the warping function W according with (18):

$$W(x, y) = A\left(\int_{x_0}^{x} \partial_y\Phi_0 dx - \int_0^y \partial_x\Phi_0 dy\right) - xy.$$

Again the real blade contour is given by the array of points (about 30), and the functions $y_t(x)$, $y_l(x)$ are defined in Mathcad by the regression (the built-in functions regress and interp).

In the projections from the vector Eq. (15), we obtain the system

$$
\begin{aligned}
&Q'_x = -q_x + \rho \ddot{u}_x, \quad Q'_y = -q_y + \rho \ddot{u}_y, \quad Q'_z = -q_z + \rho \ddot{u}_z, \\
&M'_x = Q_y - m_x + G_x \ddot{\theta}_x + G_{xy} \ddot{\theta}_y, \\
&M'_y = -Q_x - m_y + G_y \ddot{\theta}_y + G_{xy} \ddot{\theta}_x, \\
&M'_z = -m_z + G_z \ddot{\theta}_z, \\
&\theta'_x = A_x M_x + A_{xy} M_y, \quad \theta'_y = A_y M_y + A_{xy} M_x, \\
&\theta'_z = A_z(M_z + \eta_x Q_x + \eta_y Q_y) + C_z Q_z, \\
&u'_x = \theta_y + B_x Q_x + B_{xy} Q_y + A_z \eta_x M_z, \\
&u'_y = -\theta_x + B_y Q_y + B_{xy} Q_x + A_z \eta_y M_z, \\
&u'_z = B_z Q_z + C_z M_z.
\end{aligned}
\tag{20}
$$

Considering the harmonic oscillations with the frequency λ, variables are changing according to the law $u_x(z,t) = u_x(z) \sin \lambda t$, $\ddot{u}_x = -\lambda^2 u_x$, and for the amplitudes we obtain the system of ODE of twelfth order.

The coefficients of these equations contain the components of vectors and tensors rotating with the angular velocity Ω when the coordinate z is increased. The formulas of the components transformation are:

$$
\begin{aligned}
&\eta_x = \eta_1 \cos \varphi - \eta_2 \sin \varphi, \quad \eta_y = \eta_1 \sin \varphi + \eta_2 \cos \varphi; \\
&I_{xy} = \frac{I_1 - I_2}{2} \sin 2\varphi + I_{12} \cos 2\varphi, \\
&I_x = \frac{I_1 + I_2}{2} + \frac{I_1 - I_2}{2} \cos 2\varphi - I_{12} \sin 2\varphi, \\
&I_y = \frac{I_1 + I_2}{2} - \frac{I_1 - I_2}{2} \cos 2\varphi + I_{12} \sin 2\varphi.
\end{aligned}
\tag{21}
$$

Substituting (21) into (20), we obtain the difficult system of ODE with variable coefficients. However it is not hard to solve by means of computer mathematics (Mathcad). The system (20) is represented in the matrix form $Y' = D(z, Y)$, where the column Y contains 12 elements: the components of force, moment, rotation angle and displacement [the form of the function D is clear from (20)]. To the mentioned system of ODEs, boundary conditions should be added. For the cantilevered turbine blade with the fixed end $z = 0$ and another end $z = L$ free from loads, we have the conditions $Y_9 = Y_{10} = Y_{11} = Y_6 = Y_7 = Y_8 = 0$ and $Y_0 = Y_1 = Y_2 = Y_3 = Y_4 = Y_5 = 0$ correspondingly. The formulated boundary value problem is solved by the shooting method with the built-in functions sbval and rkfixed.

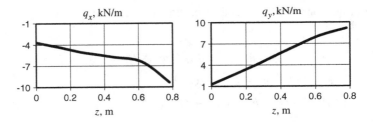

Fig. 4 The applied loads

According to the described above theory, the calculations of the real blade are done with the same parameters as above. The distributions of applied loads are shown in Fig. 4.

The calculated deflection components u_x, u_y, the angle of rotation θ_z and the axial displacements u_z are in Fig. 5. Taking into account the latter can be important [13], not only because of the centrifugal force.

For quantifying the role of the correction factors, the calculations are done without taking into account shear and cross links in the elasticity relations. Transverse forces, bending moments and deflections do not change practically (less than 0.5 %), but axial displacements and rotations vanish.

Solving of the mode analysis problems with calculation of the natural frequencies and the normal modes of free oscillations is not difficult. Here the frequency is considered as the additional variable (thirteenth): $\lambda^2 = Y_{12}, D_{12} = 0$.

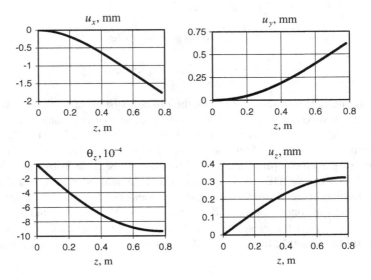

Fig. 5 The displacements and the rotations

Conclusion

So, consideration of turbine blades as rods is still advisable if there are no local peculiarities of the three-dimensional stress-strain state. New possibilities of computer mathematics allow us to calculate even the full inhomogeneous rod model with all of its compliances. Herewith in many cases ideas about the bending in two planes are enough.

Acknowledgements This work is carried out in the framework of the state task of the Ministry of Education and Science of the Russian Federation (Project No. 933-2014, 1972-2014).

References

1. Ilchenko BV, Gizzatullin RZ, Yarullin RR (2011) Stress-strain fields for blades of turbine K-210-130 under operation loading. Treatises Akademenergo (Trudy Akademenergo) 3:74–81 (in Russian)
2. Magerramova LA (2013) Increasing resource of the aviation gas turbine blades by calculation methods. Herald Mosc Aviat Inst (Vestnik Moskovskogo Aviatsionnogo Instituta) 20(1):58–70 (in Russian)
3. Melnikova GV, Shorr BF, Salnikov AV, Nigmatullin RZ (2014) Automated dynamic optimization of the blades of the gas turbine engines. Herald Mosc Aviat Inst (Vestnik Moskovskogo Aviatsionnogo Instituta) 21(1):76–85 (in Russian)
4. Trushin VA, Chechulin AYu (2012) Mathematical model of calculation of radial clearance between the working vanes and turbine housing. Herald Ufa State Aviat. Techn. Univ. (Vestnik UGATU) 16(2):82–86 (in Russian)
5. Bloch HP, Singh MP (2009) Steam turbines: design, applications and relating. McGraw-Hill, New York, p 433
6. Leyzerovich AS (2005) Wet-steam turbines for nuclear power plants. PennWell, USA, p 481
7. Brondsted P, Nijssen RPL (2013) Advances in wind turbine blade: design and materials. Woodhead Publishing Limited, Cambridge, p 484
8. Birger IA, Shorr BF et al (1981) Dynamics of aviation gas turbine engines (Dinamika aviatsionnykh gazoturbinnykh dvigatelei). Mashinostroyenie, Moscow, p 232 (in Russian)
9. Dimentberg FM, Kolesnikov, KS (1980) Vibrations in technics. Oscillations of machines, constructions and their elements (Vibratsiya v tehnike. Kolebaniya mashin, konstruktsiy i ih elementov) Mashinostroyenie, Moscow, p 544 (in Russian)
10. Levin AV, Borishanskiy KN, Konson ED (1981) Strength and vibration of blades and disks of steam turbines (Prochnost i vibratsiya lopatok i diskov parovyih turbin). Mashinostroyenie, Leningrad, p 710 (in Russian)
11. Leonov VP, Schastlivaya IA, Igolkina TN et al (2014) Application of finite element method for simulation of stress-strain state in manufacturing of long turbine blades made of high-strength titanium alloys. Inorg Mat Appl Res 5(6):578–586. doi:10.1134/S2075113314060069
12. Balakshin OB, Kukharenko BG, Khorikov AA (2007) Independent component analysis of oscillations under rotor blade flutter. Eng Autom Probl 4:77–83
13. Ganiev RF, Balakshin OB, Kukharenko BG (2009) On the occurrence of self-synchronization of autooscillations of turbocompressor rotor blades. J Mach Manuf Reliab 38(6):535–541. doi:10.3103/S105261880906003X

14. Bezjazychny VF, Ganzen MA (2011) Estimation of blade-to-disk attachments deformation in turbomachines. Handbook. Eng J 4:18–22 (in Russian)
15. Korihin NV (2009) Study of stress state of lock joint of ceramic blade of promising gas turbine. Russ J Heavy Mach 1:8–11 (in Russian)
16. Eliseev VV, Moskalets AA, Oborin EA (2015) Deformation analysis of turbine blades based on complete one-dimensional model. Russ J Heavy Mach 5:35–38 (in Russian)
17. Eliseev VV, Moskalets AA, Oborin EA (2015) Applying of lagrange equations to calculation of turbine blade vibration. Handbook. An Eng J 8:21–24. doi:10.14489/hb.2015.08. (in Russian)
18. Eliseev VV (2003) Mechanics of elastic bodies. St. Petersburg State Polytechn. University Publishing House, St Petersburg, p 336 (in Russian)
19. Kiryanov DV (2007) Mathcad 14. BHV-Peterburg, St Petersburg, p 704 (in Russian)
20. Dzhanelidze GYu (1946) Relations of Kirchhoff for naturally twisted rods and their applications (Sootnosheniya Kirchhoffa dlya estestvenno-zakruchennyih sterzhney i ih prilozheniya). Treatises Leningrad Polytechn. Univ. (Trudy Leningradskogo Politechn. Univ.) 1:23–32 (in Russian)
21. Vorobev YuS, Shorr BF (1983) Theory of twisted rods (Teoriya zakruchennyih sterzhney). Naukova Dumka, Kiev, p 188 (in Russian)
22. Eliseev VV (1991) Saint-Venant problem and elastic moduli for rods with curvature and twisting (Zadacha Saint-Venanta i uprugie moduli dlya sterzhney s kriviznoy i krucheniem). Mech. Solids. A Journal of Russian Acad. of Science (Izvestiya AN SSSR, MTT) 2:167–176 (in Russian)
23. Iesan D (2008) Classical and generalized models of elastic rods. Chapman & Hall, London, p 349. doi:10.1201/9781420086508
24. Khanh CL (1999) Vibrations of shells and rods. Springer, Berlin, p 423. doi:10.1007/978-3-642-59911-8
25. Rubin MB (2000) Cosserat theories: shells, rods and points. Springer, Berlin, p 488. doi:10.1007/978-94-015-9379-3
26. Lurie AI (2005) Theory of elasticity. Springer, Berlin, p 1038. doi:10.1007/978-3-540-26455-2

Stationary Oscillation in Two-Mass Machine Aggregate with Universal-Joint Drive

Vassil Zlatanov

Abstract The article is devoted to questions of stationary oscillation and dynamic loading of a two-mass machine aggregate with a linearly elastic intermediate shaft as a universal-joint drive. One method of analytical dynamic research on the basis of methods of nonlinear mechanics is presented.

Keywords Machine aggregate · Universal-joint drive · Lindstedt-Poincaré method

Introduction

Scientific and technical progress is inconceivable without creation of a highly effective technique, tested mechanisms and their improvement. Mechanical systems with a universal-joint drive are characterized by big load ability, compactness, and the opportunity to pass a rotation torque with an angle between shafts to 45° and at high speeds. In universal-joint drives the principle of function of the Hook joint is used. Many works are devoted to mechanics of the Hook joint's mechanics [1, 4, 8, 11–13, 18]. The interest in mechanical systems with universal-joint drive because of their "rich" and interesting mechanical properties, remains high [2, 3, 5–7, 9, 10, 14–17, 19]. Improvement of such mechanical functions causes a necessity for further development of methods of calculation, specification and improvement of the existing methods.

In this work the method of analytical dynamic research of the two-mass machine aggregate with universal-joint drive is presented, considering elasticity of its intermediate shaft.

V. Zlatanov (✉)
University of Food Technologies, Plovdiv, Bulgaria
e-mail: vassilzlatanov@mail.ru

© Springer International Publishing Switzerland 2016
A. Evgrafov (ed.), *Advances in Mechanical Engineering*,
Lecture Notes in Mechanical Engineering, DOI 10.1007/978-3-319-29579-4_11

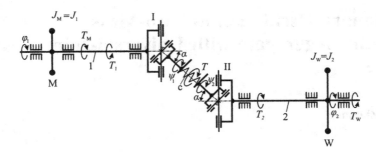

Fig. 1 Dynamic model. *1* and *2*—the input and output shaft; *I* and *II*—Hook's joints connecting shaft *1* and *2* with an intermediate shaft

Dynamic Model

The dynamic model is presented in Fig. 1. Engine M is an electric motor with constant mass inertial moment of a rotor $J_M = J_1 = \text{const}$ and the motor torque moment, changing under the linear law $T_M = a - b\omega_1$, where constants a and b are also defined by specifications of the electric motor, and ω_1 is the angular speed of a rotor.

The universal-joint drive is considered as a plane two-joint system. In the dynamic model we find: 1 and 2—respectively the input and output shaft; I and II—Hook's joints connecting shaft 1 and 2 with an intermediate shaft; α—an angle between geometrical axes of the 1st, 2nd and the intermediate shaft; φ_1 and φ_2—angles of rotation of shafts of the drive. The intermediate shaft is considered as an elastic body with elasticity coefficient c; ψ_1 and ψ_2—rotation angles of the intermediate shaft's fork. Other elements of universal-joint drive are considered as a rigid body.

The working machine W—rotational type with the constant moment of inertia $J_W = J_2$ reduced to a shaft and the constant moment of resistance $T_W = c_W$.

Differential Equation of the Motion

Angles φ_i and ψ_i are also connected by dependences [1]:

$$\text{tg}\varphi_i = \cos\alpha\,\text{tg}\psi_i \quad (i = 1, 2), \tag{1}$$

and the ratio of the two couples forks are defined from expressions

$$\delta_i = \frac{\psi_i}{\varphi_i} = \frac{\cos \alpha}{1 - \sin^2 \alpha \cos^2 \varphi_i} \quad (i = 1, 2). \tag{2}$$

Using the method of uncertain coefficients from "Eq. 2" we obtain [17]:

$$\delta_i = 1 + \sum_{k=1}^{\infty} 2\lambda^k \cos(2k\varphi_i) \quad (i = 1, 2), \tag{3}$$

and after integration:

$$\psi_i = \varphi_i + \sum_{k=1}^{\infty} \frac{(\lambda)^k}{k} \sin(2k\varphi_i) \quad (i = 1, 2). \tag{4}$$

In "Eq. 3" and "Eq. 4" it is denoted as:

$$\lambda = \text{tg}^2 \frac{\alpha}{2}, \tag{5}$$

which was a small quantity $\left(\lambda < 1, \alpha < \frac{\pi}{4}\right)$ and as a result we receive:

$$\delta_i = 1 + 2\lambda \cos(2\varphi_i), \quad \psi_i = \varphi_i + \lambda \sin(2\varphi_i) \quad (i = 1, 2). \tag{6}$$

Rotational torques T_i $(i = 1, 2)$ are expressed by means of a rotational torque T of an intermediate shaft:

$$T_i = T\delta_i \quad (i = 1, 2), \tag{7}$$

which is a consequence of the assumption of ideality of constraints, and a rotational torque T is accepted to be proportional to deformation of a shaft:

$$T = c(\psi_1 - \psi_2). \tag{8}$$

Replacing "Eq. 6" and "Eq. 8" in "Eq. 7", neglecting the members containing λ in degrees above the first, we receive:

$$T_i = c\{(\varphi_1 - \varphi_2) + 2\lambda[(\varphi_1 - \varphi_2)\cos(2\varphi_i) + \sin(\varphi_1 - \varphi_2)\cos(\varphi_1 + \varphi_2)]\}. \tag{9}$$

The differential equations of the motion of the two rotation masses have an appearance:

$$\begin{aligned}
J_1 \ddot{\varphi}_1 &= a - b\dot{\varphi}_1 - c\{(\varphi_1 - \varphi_2) + 2\lambda[(\varphi_1 - \varphi_2)\cos(2\varphi_1) \\
&\quad + \sin(\varphi_1 - \varphi_2)\cos(\varphi_1 + \varphi_2)]\}, \\
J_2 \ddot{\varphi}_2 &= -c_w + c\{(\varphi_1 - \varphi_2) + 2\lambda[(\varphi_1 - \varphi_2)\cos(2\varphi_2) \\
&\quad + \sin(\varphi_1 - \varphi_2)\cos(\varphi_1 + \varphi_2)]\}.
\end{aligned} \tag{10}$$

Solution of the Differential Equations

To solve the system of equations "Eq. 10", at the beginning we summarize and then we subtract as previously having multiplied J_2 and J_1:

$$J_1\ddot{\varphi}_1 + J_2\ddot{\varphi}_2 = T_M - T_W + 2c\lambda(\varphi_1 - \varphi_2)(\cos(2\varphi_2) - \cos(2\varphi_1)), \tag{11}$$

$$\begin{aligned}J_1 J_2(\ddot{\varphi}_1 - \ddot{\varphi}_2) = {} & J_2 T_M + J_1 T_W - c(\varphi_1 - \varphi_2)(J_1 + J_2) \\ & - 2c\lambda(\varphi_1 - \varphi_2)(J_2\cos(2\varphi_1) + J_1\cos(2\varphi_2)) \\ & - 2c\lambda(J_1 + J_2)\sin(\varphi_1 - \varphi_2)\cos(\varphi_1 + \varphi_2).\end{aligned} \tag{12}$$

We introduce the normal coordinates φ and θ:

$$\varphi = \varphi_1 - \varphi_2, \quad \theta = \frac{J_1\varphi_1 + J_2\varphi_2}{J_1 + J_2}, \tag{13}$$

from where for the previous coordinates we obtain:

$$\varphi_1 = \theta + \frac{J_2}{J_1 + J_2}\varphi, \quad \varphi_2 = \theta - \frac{J_1}{J_1 + J_2}\varphi. \tag{14}$$

The object of research—the stationary mode of the motion of the machine aggregate, i.e. $T_M = T_W$ or $a - b\omega_{cm} = c_W$, from where is defined stationary angular velocity $\omega'_{cm} = \frac{a - c_W}{b}$ and considering "Eq. 13", the system from "Eq. 11" and "Eq. 12" obtain the form:

$$(J_1 + J_2)\ddot{\theta} = 2c\lambda\varphi[\cos(2\varphi_2) - \cos(2\varphi_1)], \tag{15}$$

$$\begin{aligned}J_1 J_2\ddot{\varphi} + c\varphi(J_1 + J_2) = {} & c_W(J_1 + J_2) - 2c\lambda\varphi[J_2\cos(2\varphi_1) + J_1\cos(2\varphi_2)] \\ & - 2c\lambda(J_1 + J_2)\sin\varphi\cos(\varphi_1 + \varphi_2).\end{aligned} \tag{16}$$

The trigonometric functions on the right-hand side of "Eq. 15" are represented by means of the following power series expansion:

$$\cos(2\varphi_1) - \cos(2\varphi_2) = \sum_{n=0}^{\infty}(-1)^n\frac{\left[(2\varphi_2)^{2n} - (2\varphi_1)^{2n}\right]}{2n!}. \tag{17}$$

We consider that for the angles φ_i ($i = 1, 2$) the conditions $2\varphi_i < 1$ are completed. Therefore we use only the first two terms of the series of "Eq. 17" and as we take into account "Eq. 13" and "Eq. 14" we obtain:

$$\cos(2\varphi_1) - \cos(2\varphi_2) \approx 2(\varphi_1 - \varphi_2)(\varphi_1 + \varphi_2) = 4\varphi\theta + \frac{2(J_2 - J_1)}{J_1 + J_2}\varphi^2. \tag{18}$$

Replacing "Eq. 18" in "Eq. 15" it turns out that

$$(J_1 + J_2)\ddot{\theta} = \lambda 4c\varphi^2 \left[2\theta + \frac{(J_2 - J_1)}{J_1 + J_2}\varphi\right],$$

and we take into account that $\frac{(J_2 - J_1)}{J_1 + J_2} < 1$ and $\varphi \ll 1$, so we can neglect the second addend and finally the differential equation of "Eq. 15" obtains the form:

$$(J_1 + J_2)\ddot{\theta} = \lambda(8c\theta\varphi^2), \qquad (19)$$

where λ is a small positive parameter.

The trigonometric functions on the right-hand side of the expression of "Eq. 16", considering the received restrictions for angels φ_i ($i = 1, 2$), are represented as follows:

$$\sin\varphi \approx \varphi, \cos(2\varphi_1) = \cos\left(2\theta + 2\frac{J_2}{J_1 + J_2}\varphi\right) \approx \cos(2\theta), \cos(2\varphi_2)$$

$$= \cos\left(2\theta - 2\frac{J_1}{J_1 + J_2}\varphi\right) \approx \cos(2\theta), \cos(\varphi_1 + \varphi_2)$$

$$= \cos\left(2\theta + \frac{J_2 - J_1}{J_1 + J_2}\varphi\right) \approx \cos(2\theta),$$

and since $\varphi \ll 1, \frac{J_i}{J_1 + J_2} < 1, \frac{J_2 - J_1}{J_1 + J_2} < 1$ ($i = 1, 2$) the differential equation of "Eq. 16" appears in the following canonical form:

$$\ddot{\varphi} + c\frac{J_1 + J_2}{J_1 J_2}\varphi = c_W\frac{J_1 + J_2}{J_1 J_2} - \lambda 4c\frac{J_1 + J_2}{J_1 J_2}\cos(2\theta). \qquad (20)$$

We denote $k^2 = c\frac{J_1 + J_2}{J_1 J_2}$ and with the introduction of a new variable through equality:

$$\eta = \varphi - \frac{c_W}{k^2}\frac{J_1 + J_2}{J_1 J_2} = \varphi - \frac{c_W}{c} = \varphi - \varphi_{cm}, \qquad (21)$$

the system of the differential equations of "Eqs. 19–20" finally assumes the form:

$$(J_1 + J_2)\ddot{\theta} = \lambda(b_{11}\theta + b_{12}\eta\theta + b_{13}\eta^2\theta),$$
$$\ddot{\eta} + k^2\eta = -\lambda 4k^2\cos(2\theta), \qquad (22)$$

where λ is a small positive parameter; $b_{11} = 8c\varphi_{cm}^2; b_{12} = 16c\varphi_{cm}; b_{16} = 8c$, and φ_{cm} can be considered as an angle of torsion of an intermediate shaft with the assumption that in its two edges the moments are T_M and T_W.

The solution of the system is found by means of the method of small parameter (Poincaré), Lyapunov–Lindstedt's method in a combination with the method of stage-by-stage integration. Depending on the considered conditions of the change of angles φ_i ($i = 1, 2$) time is broken into stages. Knowing the initial conditions of the motion, it is possible to integrate these equations at the first stage and to define $\theta(t)$, $\eta(t)$, $\dot{\theta}(t)$, $\dot{\eta}(t)$ in a final time point. These final values are initial for the following stage.

The solution of expression of "Eq. 22" in compliance with the requirements of the used methods, accurate to the first power of the small parameter is the following:

$$\theta(t) = \sigma_0(t) + \lambda\sigma_1(t), \quad \eta(t) = \eta_0(t) + \lambda\eta_1(t), \tag{23}$$

where $p^2 = k^2 + \lambda h_1$, and h_1—the constant which is subject to definition.

We equate to zero coefficients at various powers of λ, and the equations turn out:

$$(J_1 + J_2)\ddot{\sigma}_0 = 0, \quad \ddot{\eta}_0 + k^2\eta_0 = 0;$$
$$(J_1 + J_2)\ddot{\sigma}_1 = b_{11}\sigma_0 + b_{12}\eta_0\sigma_0 + b_{13}\eta_0^2\sigma_0, \quad \ddot{\eta}_1 + k^2\eta_1 = h_1\eta_0 - 4k^2\cos(2\sigma_0). \tag{24}$$

We will search for the decision having in mind the conditions:

$$t = 0, \eta(0) = \varphi(0) - \varphi_{cm} = \varphi_0 - \varphi_{cm} = \varphi_1(0) - \varphi_2(0) - \varphi_{cm},$$
$$\dot{\eta}(0) = \dot{\varphi}(0) = \dot{\varphi}_0 = \dot{\varphi}_1(0) - \dot{\varphi}_2(0),$$
$$\sigma(0) = \theta_0 = \frac{J_1\varphi_1(0) + J_2\varphi_2(0)}{J_1 + J_2}, \quad \dot{\sigma}(0) = \omega_{cm} = \frac{J_1\dot{\varphi}_1(0) + J_2\dot{\varphi}_2(0)}{J_1 + J_2}. \tag{25}$$

These conditions are satisfied if functions $\sigma_0(t), \sigma_1(t), \eta_0(t), \eta_1(t)$ are found so that

$$\eta_0(0) = \varphi_0 - \varphi_{cm}, \dot{\eta}_0(0) = \dot{\varphi}_0, \eta_1(0) = 0, \dot{\eta}_1(0) = 0;$$
$$\sigma_0(0) = \theta_0, \dot{\sigma}_0(0) = \omega_{cm}, \sigma_1(0) = 0, \dot{\sigma}_1(0) = 0. \tag{26}$$

From the first and the second equations of "Eq. 24", with the initial conditions of "Eq. 26" for normal coordinates φ, θ and for the initial variables φ_1, φ_2 we find:

$$\theta = \sigma_0 = \theta_0 + \omega_{cm}t, \varphi = \eta_0 + \varphi_{cm} = \varphi_{cm} + A\cos(kt - \alpha),$$
$$\varphi_1 = \left(\theta_0 + \frac{J_2}{J_1 + J_2}\varphi_{cm}\right) + \omega_{cm}t + A_1\cos(kt - \alpha),$$
$$\varphi_2 = \left(\theta_0 - \frac{J_1}{J_1 + J_2}\varphi_{cm}\right) + \omega_{cm}t - A_2\cos(kt - \alpha), \tag{27}$$

where $A = \sqrt{(\varphi_0 - \varphi_{cm})^2 + \left(\frac{\varphi_0}{k}\right)^2}$, $\mathrm{tg}\alpha = \frac{\varphi_0}{k(\varphi_0 - \varphi_{cm})}$, $A_1 = \frac{J_2}{J_1 + J_2}A$, $A_2 = \frac{J_1}{J_1 + J_2}A$.

For the torque loading a shaft in zero approach can be presented as follows:

$$
\begin{aligned}
T &= c(\psi_1 - \psi_2) = c[(\varphi_1 - \varphi_2) + \lambda(\sin 2\varphi_1 - \sin 2\varphi_2)] \\
&\approx c\varphi[1 + 2\lambda \cos 2\theta] = [c_W + cA\cos(kt - \alpha)][1 + 2\lambda\cos(2\theta_0 + 2\omega_{CT}t)].
\end{aligned}
\tag{28}
$$

We have a constant component c_W on which a harmoniously changing component with a frequency k and amplitude A as in the two-mass unit with a linearly elastic shaft between the engine and the working machine. The harmonious component of the second addend in the last expression of "Eq. 28" is connected with the use of the universal-joint drive in the dynamic model.

After the replacement of expression of "Eq. 27" in the fourth equation on "Eq. 24" from the condition of absence of the secular term, it turns out: $h_1 = 0 \Rightarrow p^2 = k^2$. As the initial conditions of "Eq. 26" are equal to zero, the function $\eta_1(t)$ is defined by Duhamel's integral:

$$
\eta_1 = \frac{1}{k} \int_0^t \sin k(t - \tau)\left[-4k^2\cos(2\theta_0 + 2\omega_{cm}\tau)\right] d\tau.
\tag{29}
$$

The normal coordinate φ, is determined after the solution of the second equation in "Eq. 23", and as a first approximation, is received in the form:

$$
\begin{aligned}
\varphi = \varphi_{CT} &+ A\cos(kt - \alpha) \\
&+ \lambda\frac{4k^2}{k^2 - 4\omega_{CT}^2}\{-\cos(2\theta_0)[\cos(2\omega_{CT}t) - \cos(kt)] \\
&+ \sin(2\theta_0)\left[\sin(2\omega_{CT}t) - \frac{2\omega_{CT}}{k}\sin(kt)\right]\}.
\end{aligned}
\tag{30}
$$

We replace "Eq. 27" in the third equation of "Eq. 24", considering the zero initial conditions of "Eq. 26", using Duhamel's integral for the definition σ_1:

$$
\begin{aligned}
\sigma_1 = \frac{1}{J_1 + J_2} \int_0^t (t - \tau)&[b_{11}(\theta_0 + \omega_{CT}\tau) + b_{12}(\theta_0 + \omega_{CT}\tau)A\cos(k\tau - \alpha) \\
&+ b_{13}(\theta_0 + \omega_{CT}\tau)A^2\cos^2(k\tau - \alpha)]d\tau,
\end{aligned}
\tag{31}
$$

and after transformations, for the normal coordinate θ in a first approximation we receive:

$$
\begin{aligned}
\theta = {} & \theta_0 + \omega_{CT}t + \lambda \frac{1}{J_1 + J_2} \left\{ 4c\varphi_{CT}^2 \left(\theta_0 + \frac{\omega_{CT}}{3}t \right) t^2 \right. \\
& + \frac{t}{k^2} 2c\omega_{CT} \left[-8\varphi_{cm} \left((\varphi_0 - \varphi_{CT})\cos(kt) + \frac{\dot{\varphi}_0}{k}\sin(kt) \right) - \left(\frac{1}{2} \left((\varphi_0 - \varphi_{CT})^2 \right. \right. \right. \\
& + \left(\frac{\dot{\varphi}_0}{k} \right)^2 \right) \cos(2kt) + (\varphi_0 - \varphi_{CT}) \frac{\dot{\varphi}_0}{k} \sin(2kt) \Big) \Big] + 16c\varphi_{CT} \Big[-\Big(\theta_0 \big(\varphi_0 - \varphi_{CT} \big) \\
& + 2\omega_{CT} \frac{\dot{\varphi}_0}{k} \Big) \cos(kt) + \left(-\theta_0 \frac{\dot{\varphi}_0}{k} + 2\frac{\omega_{CT}(\varphi_0 - \varphi_{CT})}{k} \right) \sin(kt) \Big] \\
& + 2c \left[-\left(0.5\theta_0 \left((\varphi_0 - \varphi_{CT})^2 + \left(\frac{\dot{\varphi}_0}{k} \right)^2 \right) + \frac{\omega_{CT}(\varphi_0 - \varphi_{CT})\dot{\varphi}_0}{k^2} \right) \cos(2kt) \right. \\
& + \left. \left. \left(0.5\omega_{CT} \left((\varphi_0 - \varphi_{CT})^2 + \left(\frac{\dot{\varphi}_0}{k} \right)^2 \right) - \theta_0(\varphi_0 - \varphi_{CT})\frac{\dot{\varphi}_0}{k} \right) \sin(2kt) \right] \right\}.
\end{aligned}
$$

$$(32)$$

The received functions in "Eq. 30" and "Eq. 32" allow the definition of laws of motion of the machine aggregate through "Eq. 14" with precision to the first approximation. The torque loading an intermediate shaft at the considered stage of stage-by-stage integration is obtained from "Eq. 8" or "Eq. 28".

Conclusion

In this work the dynamic model of a mechanical system with universal-joint drive as a two-mass machine aggregate was presented. The universal-joint drive was considered as a plane two-joint system. The intermediate shaft of a universal-joint drive was presented as a linearly elastic body. The laws of motion of the machine aggregate and torque loading an intermediate shaft by means of the method of small parameter (Poincaré), Lyapunov–Lindstedt's method in a combination with the method of stage-by-stage integration was obtained.

The presented analytical method completes the famous methods in the theory of dynamic research of mechanical system with universal-joint drive.

References

1. Artobolevskij II (1988) Theory of mechanisms and machines. Nauka, Moscow (rus.)
2. Asokantan SF, Meehan PA (2000) Non-linear vibration of a torsional system driven by a Hooke's joint. J Sound Vib 233(2):297–310. doi:10.1006/jsvi.1999.2802
3. Bass KM, Plahotnik VV, Krivda VV (2012) Mathematical model of fluctuations in rectilinear motion of a dump truck. In: Modern engineering: science and education, 2nd international scientific and practical conference, Saint Petersburg, Russia, June 2012. Lecture notes in

modern engineering: science and education. Publishing House of Polytechnical University, Saint Petersburg, pp 161–165 (rus.)

4. Buchvarov SN, Pisarev AM, Cheshankov BI (1967) Mechanics of Cardan motion of a rigid body. Ann VTUS Appl Mech III(II):195–205 (bul.)

5. Buchvarov SN, Zlatanov VD, Delcheva SN (2005) Dynamics of machine aggregate with elastic shaft and actuating mechanism with quadratic characteristic curve (part I). Theory Mech Mach 5(3):24–34. http://tmm.spbstu.ru/5/zlatanov_2005.pdf (rus.)

6. Buchvarov SN, Zlatanov VD, Delcheva SN (2006) Dynamics of machine aggregate with elastic shaft and actuating mechanism with quadratic characteristic curve (part II). Theory Mech Mach 7(4):72–80. http://tmm.spbstu.ru/7/buchvarov-2.pdf (rus.)

7. Bulut G, Parlar Z (2011) Dynamic stability of a shaft system connected through a Hooke's joint. Mech Mach Theory 46(11):1689–1695. doi:10.1016/j.mechmachtheory.2011.06.012

8. Duditza F (1973) Kardangelenkgetnebe und ihre Anwendungen. VDI-Verlag, Dusseldorf (ger)

9. Efremov LV (2013) Torsional vibrations drive power unit with elastic coupling type LMD. In: Modern engineering: science and education, 3rd international scientific and practical conference, Saint Petersburg, Russia, June 2013. Lecture notes in modern engineering: science and education. Publishing House of Polytechnical University, Saint Petersburg, pp 318–325 (rus.)

10. Kozhevnikov SN, Perfileev PD (1974) Research of torsional oscillations of transmission with several universal-joint drives. Theory Mech Mach 16:32–39 (rus.)

11. Kozhevnikov SN, Perfileev PD (1978) Universal-joint drives. Technika, Kiev (rus.)

12. Lojtsyanskij LG, Lurie AI (1983) Course of theoretical mechanics, T.II. Nauka, Moscow (rus.)

13. Malahovskij IE, Lapin AA, Venedeev NK (1962) Universal-joint drives. Mashgiz, Moscow (rus.)

14. Morozov BI, Pchelin IK, Hachaturov AA (1965) Task for torsional oscillations of transmission with universal-joint drive. Scientific works of NAMI, vol 74, pp 19–23 (rus.)

15. Perfileev PD (1972) Torsional oscillations in the transmission with universal-joint drive. Scientific works of IPI, vol 75, pp 265–273 (rus.)

16. Pisarev AM (1969) Dynamics of a single universal-joint drive with elastic links and an engine. Proc Bul Acad Sci 22(3):241–244 (rus., bul.)

17. Pisarev AM, Cheshankov BI, Buchvarov SN (1968) Torsional oscillations in universal-joint shafts. Ann VTUS Appl Mech IV(I):13–22 (bul.)

18. Poliakov AA (1999) Bases of strength and dynamics of one class nonlinear spatial joint-shell systems. UGTU, Yekaterinburg (rus.)

19. Tomchina OM, Reznichenko VV, Terenteva OV (2013) Dynamics of two-rotor vibration unit with elastic cardan shafts. In: Modern engineering: science and education, 3rd international scientific and practical conference, Saint Petersburg, Russia, June 2013. Lecture notes in modern engineering: science and education. Publishing House of Polytechnical University, Saint Petersburg, pp 841–850 (rus.)

The Vibrations of Reservoirs and Cylindrical Supports of Hydro Technical Constructions Partially Submerged into the Liquid

George V. Filippenko

Abstract We consider here the free oscillations of a system of concentric cylindrical shells partially submerged in a fluid and vertically attached to a reservoir bottom. The statement of the problem is a rigorous one. The dispersion equation and reflection coefficients are found on the basis of an exact analytical solution of the submerged part of a single cylindrical shell. The generalizations of this method are discussed. We use a method of impedances for constructing dispersion equations and generalized displacements and forces. Look-alike systems can model the supports of hydro technical constructions, oil rigs, wind electro stations, standing on a shelf, reservoirs with fluid, etc.

Keywords Cylindrical shell · Free vibrations of a shell · Input impedance of a shell · Boundary contact problem of acoustics

Introduction and Statement of the Problem

Cylindrical shells (sometimes multilayered) are used in construction of supports of hydraulic engineering constructions, oil platforms, wind electric powers, construction on a shelf and also reservoirs containing a liquid, etc. Calculation of these complicated systems demands major computing resources. Therefore, consideration of simpler model problems which have exact analytical solutions [1–14] is crucial. It is possible to analytically explore the main effects on these models and also to use them as test problems for computing packages. The solution is based on the general scheme stated in [7] for one shell, and also on the technology of input impedances [15].

G.V. Filippenko (✉)
Institute of Problems of Mechanical Engineering, V.O, Bolshoj pr, 61,
St Petersburg 199178, Russia
e-mail: g.filippenko@gmail.com; g.filippenko@spbu.ru

G.V. Filippenko
St.Petersburg State University, 7-9, Universitetskaya nab, St.Petersburg 199034, Russia

© Springer International Publishing Switzerland 2016
A. Evgrafov (ed.), *Advances in Mechanical Engineering*,
Lecture Notes in Mechanical Engineering, DOI 10.1007/978-3-319-29579-4_12

115

Let us start from a cylindrical system of coordinates where axis $0z$ coincides with the axis of a cylinder and choose the variables that describe the system: the acoustic pressure $P(x, y)$ in the liquid and the displacement vector of the shell $\mathbf{u} = (u_t, u_z, u_n)^T$ (T—is a badge of transposing). The liquid is supposed to be ideal and compressible and it is described by the Helmholtz equation,

$$(\Delta + k^2)P(r, \varphi, z) = 0; k = \frac{\omega}{c}, R < r < +\infty, 0 \le \varphi < 2\pi, 0 < z < H.$$

The surface $z = H$ of the liquid layer is free $P(r, \varphi, z)|_{z=H} = 0$ and the bottom of the reservoir $z = 0$ is rigid $\frac{\partial P(r,\varphi,z)}{\partial z}\Big|_{z=0} = 0$ ($R < r < +\infty$, $0 \le \varphi < 2\pi$).

The empty cylindrical shell is vertically attached to the bottom of the reservoir and partially protrudes ($H < z < l$) above the surface of the liquid. The source of an acoustic field in a wave guide is the free vibrations of the shell.

Two relations take place on the shell—fluid boundary: the dynamic (balance of forces acting on the shell)

$$\widetilde{\mathbf{L}}^\partial \mathbf{u} = \frac{R^2 P(r, \varphi, z)|_{r=R}}{\rho c_s^2} \begin{bmatrix} (0,0,1)^T, & 0 < z < H, & 0 \le \varphi < 2\pi \\ (0,0,0)^T, & H < z < l, & 0 \le \varphi < 2\pi \end{bmatrix}$$

and kinematic (the adhesion condition) for the immersed part of the shell

$$u_n(R, \varphi, z) = \frac{1}{\rho_w \omega^2} \frac{\partial P(r, \varphi, z)}{\partial r}\Big|_{r=R+0}; \quad 0 < z < H, \ 0 \le \varphi < 2\pi$$

Here $\widetilde{\mathbf{L}}^\partial$ is the matrix differential operator [7, 16]

$$\widetilde{\mathbf{L}}^\partial \equiv w^2 \mathbf{I} + \mathbf{L}^\partial; \quad \mathbf{I} \equiv \begin{pmatrix} 1 & 0 & 0 \\ 0 & 1 & 0 \\ 0 & 0 & 1 \end{pmatrix},$$

$$\mathbf{L}^\partial \equiv \begin{pmatrix} \alpha_1[v_- \widetilde{\partial}_z^2 + \partial_\varphi^2] & v_+ + \partial_\varphi \widetilde{\partial}_z & \partial_\varphi(1 + 2\alpha^2[1 - \partial_\varphi^2 - \widetilde{\partial}_z^2]) \\ (\mathbf{L}^\partial)_{21} & \widetilde{\partial}_z^2 + v_- \partial_\varphi^2 & v\widetilde{\partial}_z \\ (\mathbf{L}^\partial)_{31} & (\mathbf{L}^\partial)_{32} & \alpha^2(2v\widetilde{\partial}_z^2 + 2\partial_\varphi^2 - 1 - [\widetilde{\partial}_z^2 + \partial_\varphi^2]^2) - 1 \end{pmatrix},$$

where $\widetilde{\partial}_z = R\partial_z$, $(\mathbf{L}^\partial)_{21} = (\mathbf{L}^\partial)_{12}$, $(\mathbf{L}^\partial)_{31} = -(\mathbf{L}^\partial)_{13}$, $(\mathbf{L}^\partial)_{32} = -(\mathbf{L}^\partial)_{23}$, $v_+ = (1+v)/2$, $v_- = (1-v)/2$, $\alpha_1 = 1 + 4\alpha^2$, $w = \omega R/c_s$. The following geometrical parameters of the shell are used: R—radius, h—thickness, l—height. Here $\alpha^2 = \frac{1}{12}\left(\frac{h}{R}\right)^2$ is the dimensionless parameter describing relative thickness of the cylindrical shell.

Properties of a material of the cylinder are characterized by E, v and ρ_s—Joung's module, Poisson coefficient and volumetric density accordingly.

The surface density of the shell ρ ($\rho = \rho_s h$) and c_s—the velocity of median surface deformation waves of the cylindrical shell $c_s = \sqrt{E/(1 - v^2)\rho}$ are introduced.

Boundary conditions are supplemented by the contact conditions at the top $z_0 = H$ and the bottom $z_0 = 0$ edges of the cylinder [12, 16]. For example in the case of a fixed edge it is the condition of absence of displacements and rotations

$$(\mathbf{U}^\partial \mathbf{u})\big|_{z=z_0} = \begin{pmatrix} 0 \\ 0 \\ 0 \\ 0 \end{pmatrix}; \quad \mathbf{U}^\partial = \begin{pmatrix} 1 & 0 & 0 \\ 0 & 1 & 0 \\ 0 & 0 & 1 \\ 0 & 0 & \partial_z \end{pmatrix}.$$

In the case of a free edge, it is the condition of absence of forces and moments

$$(\mathbf{F}^\partial \mathbf{u})\big|_{z=0} = \begin{pmatrix} 0 \\ 0 \\ 0 \\ 0 \end{pmatrix};$$

$$\mathbf{F}^\partial = \frac{\rho c^2}{R} \begin{pmatrix} -v_- \alpha_1 \widetilde{\partial}_z & -v_- \partial_\varphi & 4v_- \alpha^2 \widetilde{\partial}_z \partial_\varphi \\ -v\partial_\varphi & -\widetilde{\partial}_z & -v \\ -2\alpha^2 \widetilde{\partial}_z \partial_\varphi & 0 & \alpha^2 \left[(2 - v)\partial_\varphi^2 - v + \widetilde{\partial}_z^2 \right] \widetilde{\partial}_z \\ -2\alpha^2 Rv\partial_\varphi & 0 & \alpha^2 R \left[\widetilde{\partial}_z^2 - v(1 - \partial_\varphi^2) \right] \end{pmatrix}.$$

In the place of the cylinder entrance in the liquid $z = H$, the continuity of displacements, moments and cutting forces takes place correspondingly. It is caused by absence of the fluid's concentrated action on the shell in this place:

$$\begin{pmatrix} \mathbf{U}^\partial \\ \mathbf{F}^\partial \end{pmatrix} \mathbf{u}\bigg|_{z=H+0} - \begin{pmatrix} \mathbf{U}^\partial \\ \mathbf{F}^\partial \end{pmatrix} \mathbf{u}\bigg|_{z=H-0} = (0,0,0,0,0,0,0,0)^T.$$

We supplement boundary conditions by the condition at infinity. We take into account the waves in the liquid radiating from the shell and the waves damping as the distance from it increases.

Determination of the General Representation for Acoustic and Vibrational Fields

The exact expression for displacements of a shell can be derived only after defining a field in the medium. So we come to the boundary-contact problem for the Helmholtz equation in the liquid. The construction of the solution begins from representation of the acoustic pressure P in the liquid as an expansion in plane waves (normal modes of the waveguide): radiating from the shell and the waves damping as the distance from it increases

$$P(r, \varphi, z) = \sum_{m,n=0}^{\infty} p_{mn} \cos(m\varphi + \varphi_m^0) f_n(z) H_m^{(1)}(\lambda_n r),$$

where $f_n(z) = \cos(\mu_n z)$, $\mu_n = \frac{\pi}{H}(n + \frac{1}{2})$, $\lambda_n = \sqrt{k^2 - \mu_n^2}$, $n = 0, 1, 2, \ldots$. Here the following designations are entered: $H_m^{(1)}$—Hankel function with index m, p_{mn}—amplitudes of harmonics, φ_m^0—shift of a phase in the harmonic of number m.

The further construction of the solution is considered in [7] and is omitted for brevity. As a result we obtain the general solution for the immersed part of the shell in the form

$$\mathbf{u} = \sum_{m=0}^{\infty} \begin{pmatrix} A_m \sin(m\varphi + \varphi_m^0) \\ B_m \cos(m\varphi + \varphi_m^0) \\ C_m \cos(m\varphi + \varphi_m^0) \end{pmatrix}, \quad \mathbf{u}_m := \begin{pmatrix} A_m \\ B_m \\ C_m \end{pmatrix} = \sum_{j=1}^{8} t_{jm} \left\{ \mathbf{u}_{jm}^0 + \sum_{n=0}^{\infty} \frac{\psi_{mn}^j \beta_{mn}}{h_{mn}} \mathbf{u}_{mn} \right\};$$

$$\mathbf{u}_{jm}^0 = e^{i\lambda_{jm}z} \begin{pmatrix} \eta_{jm} \\ \xi_{jm} \\ \varsigma_{jm} \end{pmatrix}, \quad \mathbf{u}_{mn} = \begin{pmatrix} \hat{l}_{31}^m \cos(\mu_n z) \\ \hat{l}_{32}^m \sin(\mu_n z) \\ \hat{l}_{33}^m \cos(\mu_n z) \end{pmatrix}$$

$$(1)$$

Let us note that $\mathbf{u}_m^0 = \sum_{j=1}^{8} t_{jm} \mathbf{u}_{jm}^0$ is the general solution of a corresponding homogeneous system. The following designations are entered:

$$\psi_{mn}^j = \frac{2}{H} \int_0^H e^{i\lambda_{jm}z} f_n(z) dz, \quad \beta_{mn} = -\frac{R^2}{\rho c_s^2} \frac{H_m^{(1)}(\lambda_n R)}{\Delta_{mn}},$$

$$h_{mn} = \frac{1}{\rho_w \omega^2} \left. \frac{\partial H_m^{(1)}(\lambda_n r)}{\partial r} \right|_{r=R+0} - \beta_{mn} \hat{l}_{33}.$$

Here t_{jm}, $(j = 1, \ldots, 8)$ are the previously unknown constants, λ_{jm}, $(j = 1, \ldots, 8)$ are the eigen numbers, and $(\eta_{jm}, \xi_{jm}, \varsigma_{jm})^T$ are the eigen vectors corresponding to the homogeneous problem of free vibrations of infinity "dry" shell, normalized to unity; Δ_{mn} the determinant of the matrix $\widetilde{\mathbf{L}}$, where $\widetilde{\mathbf{L}} \equiv w^2 \mathbf{I} + \mathbf{L}$; $\mathbf{L} \equiv \{L_{ij}\}$; $i, j = 1, 2, 3$ and

$$\mathbf{L} \equiv \begin{pmatrix} \alpha_1[v - \widetilde{\mu}_n^2 + m^2] & -v + m\widetilde{\mu}_n & -m(1 + 2\alpha^2[1 + m^2 + \widetilde{\mu}_n^2]) \\ L_{21} & \widetilde{\mu}_n^2 - v - m^2 & -v\widetilde{\mu}_n \\ L_{31} & L_{32} & -\alpha^2(1 + 2v\widetilde{\mu}_n^2 + 2m^2 + [\widetilde{\mu}_n^2 + m^2]^2) - 1 \end{pmatrix},$$

where $\widetilde{\mu}_n = R\mu_n, L_{21} = L_{12}, L_{31} = L_{13}, L_{32} = L_{23}$, and $\widehat{l}_{31}, \widehat{l}_{32}, \widehat{l}_{33}$ are the corresponding minors of the matrix $\{\widetilde{L}_{ij}\}; i,j = 1,2,3$. For example, minor $\widehat{l}_{31}$ corresponds to an element of the matrix $\widetilde{L}_{31}$.

The Cylindrical Shell Partially Filled with Liquid

We now consider the cylindrical shell partially filled with liquid. In this case the liquid is inside the cylinder and fills it up to the height H. All the basic equations and boundary conditions are the same. The scheme of the solution also remains the same. So we concentrate only on the differences.

Pressure $\widetilde{P}$ in a liquid is expressed now via Bessel functions J_m (instead of $H_m^{(1)}$),

$$\widetilde{P}(r, \varphi, z) = \sum_{m,n=0}^{\infty} \widetilde{P}_{m,n} \cos(m\varphi + \varphi_m^0) f_n(z) J_m(\widetilde{\lambda}_n r),$$

and satisfies the condition of dynamic balance with another sign

$$\mathbf{L}^{\partial}\mathbf{u} = -\frac{R^2}{\rho c_s^2}(0,0,\widetilde{P})^T; \quad 0 < z < H, \ 0 \le \varphi < 2\pi.$$

The Cylindrical Shell Partially Filled with Liquid and Surrounded by Liquid

Let us consider the case of a cylindrical shell partially filled with the liquid and surrounded by the liquid. Liquid from the both sides of the shell can be different but of the same level H. The parameters corresponding to the liquid inside of the shell will be noted by the letters with the wave over them.

The pressure $\widetilde{P}$ inside of the shell has the view (2) and pressure P outside of it has the view (1). And it satisfies the dynamic conditions

$$\mathbf{L}^{\partial}\mathbf{u} = \frac{R^2}{\rho c_s^2}(0,0,P|_{r=R+0} - \widetilde{P}|_{r=R-0})^T; \quad 0 < z < H, \ 0 \le \varphi < 2\pi \qquad (2)$$

and the adhesion condition

$$u_n = \frac{1}{\rho_w \omega^2} \frac{\partial P(r, \varphi, z)}{\partial r}\bigg|_{r=R+0} = \frac{1}{\widetilde{\rho}_w \omega^2} \frac{\partial \widetilde{P}(r, \varphi, z)}{\partial r}\bigg|_{r=R-0}; \quad 0 < z < H, \ 0 \le \varphi < 2\pi.$$

From the last equation follows the relation between expansion coefficients $\tilde{p}_{m,n}$ and $p_{m,n}$ in internal and external expansions. So the dynamic condition (2) turns into

$$\mathbf{L}^{\partial}\mathbf{u} = \frac{R^2}{\rho c_s^2} \sum_{m,n=0}^{\infty} p_{mn}G_{mn}\cos(m\varphi + \varphi_m^0)f_n(z)(0,0,1)^T,$$

$$G_{mn} = H_m^{(1)}(\lambda_n R) - J_m(\tilde{\lambda}_n R)\frac{\tilde{\rho}_w}{\rho_w}\left[\frac{\partial H_m^{(1)}(\lambda_n r)}{\partial r}\Big|_{r=R+0}\right]\left[\frac{\partial J_m(\tilde{\lambda}_n r)}{\partial r}\Big|_{r=R-0}\right]^{-1}.$$

Further the construction follows the common scheme. We only change function $H_m^{(1)}$ on G_{mn}.

The Case of Two Coinciding Cylindrical Shells with the Liquid Between Them

We consider now the generalization of the method in the case of two coinciding cylindrical shells with a liquid of the level H between them. The parameters of the external shell will be designated by the "wave" over them. Notice that the material, geometrical parameters of the cylinders, can differ as well as the boundary conditions on the butts. The pressure in the liquid now has the view

$$P(r,\varphi,z) = \sum_{m,n=0}^{\infty} \cos(m\varphi + \varphi_m^0)f_n(z)[p'_{mn}J_m(\lambda_n r) + p_{mn}N_m(\lambda_n r)].$$

Also the dynamic conditions for the internal and external shells are

$$\mathbf{L}^{\partial}\mathbf{u} = \frac{R^2}{\rho c_s^2}(0,0,P|_{r=R})^T, \quad \tilde{\mathbf{L}}^{\partial}\mathbf{u} = -\frac{\tilde{R}^2}{\tilde{\rho}\tilde{c}_s^2}(0,0,P|_{r=\tilde{R}})^T; \quad 0<z<H, \ 0\leq\varphi<2\pi$$

correspondingly. At first, according to the common scheme, the general solution $\mathbf{u}_m$ for a nonhomogeneous problem for the submerged part of an internal shell is founded on

$$\mathbf{u}_m = \mathbf{u}_m^0 + \sum_{n=0}^{\infty}\langle\mathbf{p}_{mn},\mathbf{a}_{mn}\rangle\mathbf{u}_{m,n}; \quad \mathbf{a}_{mn} = \begin{pmatrix}\alpha_{mn}\\\alpha'_{mn}\end{pmatrix}, \quad \mathbf{p}_{mn} = \begin{pmatrix}p_{mn}\\p'_{mn}\end{pmatrix}. \quad (3)$$

Here

$$\langle\mathbf{p}_{mn},\mathbf{a}_{mn}\rangle = p_{mn}\alpha_{mn} + p'_{mn}\alpha'_{mn}; \quad \alpha_{mn} = -\frac{R^2}{\rho c_s^2}\frac{N_m(\lambda_n R)}{\Delta_{mn}}, \quad \alpha'_{mn} = -\frac{R^2}{\rho c_s^2}\frac{J_m(\lambda_n R)}{\Delta_{mn}}.$$

For the displacements $\tilde{\mathbf{u}}$ of an external shell the analogous expression (3) is obtained,

$$\tilde{\mathbf{u}} = \tilde{\mathbf{u}}_m^0 + \sum_{n=0}^{\infty} \left(p_{mn}\tilde{\alpha}_{mn} + p'_{mn}\tilde{\alpha}'_{mn} \right) \tilde{\mathbf{u}}_{m,n} = \tilde{\mathbf{u}}_m^0 + \sum_{n=0}^{\infty} \langle \mathbf{p}_{mn}, \tilde{\mathbf{a}}_{mn} \rangle \tilde{\mathbf{u}}_{m,n}$$

where $\tilde{\mathbf{u}}^0$ is the general solution of a homogeneous system for the external shell. Then the adhesion condition on internal and external shells is used. As result, we obtain a linear algebraic system for two equations with respect to $p_{m,n}$ and p'_{mn}. The solution can be written in the vector form (Kramer formulas)

$$\mathbf{p}_{mn} = \frac{1}{\delta_{mn}} \sum_{j=1}^{8} \mathbf{A}_{mn}^j \mathbf{z}_{jm}; \quad \mathbf{A}_{mn}^j = \begin{pmatrix} \psi_{mn}^j \tilde{j}_{mn} & -\tilde{\psi}_{mn}^j j_{mn} \\ -\psi_{mn}^j \tilde{y}_{mn} & \tilde{\psi}_{mn}^j y_{mn} \end{pmatrix},$$

$$\mathbf{z}_{jm} = \begin{pmatrix} t_{jm} \\ \tilde{t}_{jm} \end{pmatrix}, \quad \delta_{mn} = \begin{vmatrix} y_{mn} & j_{mn} \\ \tilde{y}_{mn} & \tilde{j}_{mn} \end{vmatrix};$$

$$y_{mn} = \frac{1}{\rho_w \omega^2} \frac{\partial N_m^{(1)}(\lambda_n r)}{\partial r}\bigg|_{r=R+0} - \alpha_{mn}\widehat{l}_{33}, \quad j_{mn} = \frac{1}{\rho_w \omega^2} \frac{\partial J_m^{(1)}(\lambda_n r)}{\partial r}\bigg|_{r=R+0} - \alpha'_{mn}\widehat{l}_{33}.$$

The dependence of a solution structure via constants t_{jm} and $\tilde{t}_{jm}$ is emphasized for convenience. After substituting the expression for $\mathbf{p}_{mn}$ in expression (3) the following formula is obtained:

$$\mathbf{u}_m = \sum_{j=1}^{8} \left\{ t_{jm}\mathbf{u}_{jm}^0 + \sum_{n=0}^{\infty} \frac{1}{\delta_{mn}} \left\langle \left(\mathbf{A}_{mn}^j\right)^T \mathbf{a}_{mn}, \mathbf{z}_{jm} \right\rangle \mathbf{u}_{m,n} \right\}.$$

Hence the general solution of a nonhomogeneous system for an internal shell via 16 until now unknown constants: eight t_{jm} and eight $\tilde{t}_{jm}$ is obtained.

Notice that this method of construction of a solution for the submerged part of the shell is naturally generalized in the case of an arbitrary number of coinciding cylindrical shells (with different properties) and different liquids between them. A combination of the considered methods is needed. The only requirement is the equality of liquid levels in all volumes.

The Concept of Input Impedance for a Cylindrical Shell

The generalized vectors of velocities $\mathbf{v}$ and forces $\mathbf{f}$: $\mathbf{v} = \left(v_1, v_2, v_3, v_4, \right)^T = \mathbf{U}^\partial \mathbf{u}$, $\mathbf{f} = \left(f_1, f_2, f_3, f_4 \right)^T = \mathbf{F}^\partial \mathbf{u}$ are taken into consideration. Let us suppose that displacement $\mathbf{u}$ can be expressed in the form $\mathbf{u} = \sum_{j=1}^{4} t_j \tilde{\mathbf{u}}^j = \tilde{\mathbf{U}}\mathbf{t}$. Here vector $\mathbf{t} = (t_1, t_2, t_3, t_4)^T$ and matrix $\tilde{\mathbf{U}} = (\tilde{\mathbf{u}}^1; \tilde{\mathbf{u}}^2; \tilde{\mathbf{u}}^3; \tilde{\mathbf{u}}^4)$ with columns forming by vectors $\tilde{\mathbf{u}}^j, j = 1, 2, 3, 4$

are introduced for convenience. Hence $\mathbf{v} = \mathbf{U}^\partial \mathbf{u} = \mathbf{U}^\partial \widetilde{\mathbf{U}} \mathbf{t} \equiv \mathbf{U} \mathbf{t}$. The force $\mathbf{f} = \mathbf{F}^\partial \mathbf{u} = \mathbf{F}^\partial \widetilde{\mathbf{U}} \mathbf{t} \equiv \mathbf{F} \mathbf{t}$ is represented in a similar way.

The cross section $z = z_0$ of the shell is considered. The input impedance of the shell [15] at $z = z_0$ is defined as matrix $\mathbf{Z}$, connecting kinematic variables (vector of generalized velocities) and dynamic variables (vector of generalized forces) calculated at $z = z_0$, $\mathbf{f} = \mathbf{Z}\partial_t\mathbf{u}$. A vector of generalized velocities differs from a vector of generalized coordinates by factor $(-i\omega)$ due only to the fact that the process is stationary. This factor will be related to the matrix $\mathbf{Z}$, saving for it the previous notation i.e. $\mathbf{f} = \mathbf{Z}\mathbf{u}$. Substituting here the expressions for $\mathbf{f}$ and $\mathbf{u}$, one can obtain $\mathbf{F}\mathbf{t} = \mathbf{Z}\mathbf{U}\mathbf{t}$. Hence, due to arbitrariness in the representation of $\mathbf{t}$, we get the expression $\mathbf{Z} = \mathbf{F}\mathbf{U}^{-1}$ for impedance. We note that it is sufficient to know the decisive expressions for vector functions $\widetilde{\mathbf{u}}^j$ in order to build decisive expressions for impedance $\mathbf{Z}$. Further the method of their construction will be shown.

The Input Impedance of a Submerged Part of the Cylindrical Shell

Let us find for example the input impedance at $z = H$ of submerged part $0 < z < H$ of a cylindrical shell rigidly fixed at $z = 0$ and surrounded by liquid up to level $z = H$. The field of displacements (1) in the submerged part of the shell can be represented in the form (it is sufficient to consider the fixed harmonic on m and this index will be omitted)

$$\mathbf{u} = \sum_{j=1}^{8} t_j \mathbf{u}^j = (\mathbf{U}; \widehat{\mathbf{U}})(\mathbf{t}; \widehat{\mathbf{t}}) = \mathbf{U}\mathbf{t} + \widehat{\mathbf{U}}\widehat{\mathbf{t}}.$$

Here block matrix $(\mathbf{U}; \widehat{\mathbf{U}})$ is constructed of two blocks—matrixes $\mathbf{U}$ and $\widehat{\mathbf{U}}$ of dimension $3 * 4$, which are formed from column vectors $\mathbf{u}^j$; $\mathbf{U} = (\mathbf{u}^1; \mathbf{u}^2; \mathbf{u}^3; \mathbf{u}^4)$, $\widehat{\mathbf{U}} = (\mathbf{u}^5; \mathbf{u}^6; \mathbf{u}^7; \mathbf{u}^8)$, and vector $(\mathbf{t}; \widehat{\mathbf{t}})$—is formed from components of two vectors $\mathbf{t}$ and $\widehat{\mathbf{t}}$; $\mathbf{t} = (t_1, t_2, t_3, t_4)$, $\widehat{\mathbf{t}} = (t_5, t_6, t_7, t_8)$. For instance the condition of rigid fixing of foundation of the shell $(\mathbf{U}^\partial \mathbf{u})\big|_{z=0} = \mathbf{0}$ is considered. The linear homogeneous system of four equations via eight unknown constants t_i is obtained as the result. The remaining four constants can be expressed via four first constants (maybe after renaming) if the range of the matrix is equal to four. Hence the solution of our initial system can be expressed in the form

$$\mathbf{u} = \mathbf{U}\mathbf{t} + \widehat{\mathbf{U}}\widehat{\mathbf{t}} = \left(\mathbf{U} - \widehat{\mathbf{U}}\widehat{\mathbf{U}}_0^{-1}\mathbf{U}_0\right)\mathbf{t} \equiv \widetilde{\mathbf{U}}\mathbf{t} = \sum_{j=1}^{4} t_j \widetilde{\mathbf{u}}^j; \qquad \widetilde{\mathbf{U}} = \mathbf{U} - \widehat{\mathbf{U}}\widehat{\mathbf{U}}_0^{-1}\mathbf{U}_0,$$

$$\mathbf{U}_0 = (\mathbf{U}^\partial \mathbf{U})\big|_{z=0},$$

where $\tilde{\mathbf{u}}^j$ is the vector formed from column number j of matrix $\widetilde{\mathbf{U}}$. Further along the way show in the beginning of the previous section, the decisive value of the matrix of impedances $\mathbf{Z}$ at $z = H$ is obtained. It can be rewritten by taking into account the received representations for matrix $\widetilde{\mathbf{U}}$,

$$
\begin{aligned}
\mathbf{Z} = \mathbf{F}\mathbf{U}^{-1} &= \left[\left.\left(\mathbf{F}^\partial \widetilde{\mathbf{U}}\right)\right|_{z=H}\right]\left[\left.\left(\mathbf{U}^\partial \widetilde{\mathbf{U}}\right)\right|_{z=H}\right]^{-1} \\
&= \left[\mathbf{F}_H - \widehat{\mathbf{F}}_H \widehat{\mathbf{U}}_0^{-1}\mathbf{U}_0\right]\left[\mathbf{U}_H - \widehat{\mathbf{U}}_H \widehat{\mathbf{U}}_0^{-1}\mathbf{U}_0\right]^{-1}.
\end{aligned}
$$

Here the new (numeric) matrixes are involved: $\mathbf{F}_H = \left.(\mathbf{F}^\partial \mathbf{U})\right|_{z=H}$, $\mathbf{U}_H = \left.(\mathbf{V}^\partial \mathbf{U})\right|_{z=H}$, $\widehat{\mathbf{F}}_H = \left.(\mathbf{F}^\partial \widehat{\mathbf{U}})\right|_{z=H}$, $\widehat{\mathbf{U}}_H = \left.(\mathbf{V}^\partial \widehat{\mathbf{U}})\right|_{z=H}$. In the case of a free foundation of the cylindrical shell $z = 0$, the analogous formulas can be obtained with replacement of matrix $\mathbf{U}_0$ on matrix $\mathbf{F}_0 = \left.(\mathbf{F}^\partial \mathbf{U})\right|_{z=0}$.

The Dispersion Equation of a Cylindrical Shell Partially Submerged into Liquid in the Terms of Impedances

The protruding part $H < z < l$ of the shell is considered. Let us suppose that it is rigidly fixed at $z = l$. The input impedance of the protruding part at $z = H$ is founded, analogously taking into account the absence of contact of this part with liquid. By note «+» will be marked the obtained impedance $\mathbf{Z}_+$ of the protruding part of the shell, generalized velocities $\mathbf{v}_+$ and forces $\mathbf{f}_+$ on the boundary $z = H + 0$ acting from the upper (protruding) part of the shell, and by note «–» will be marked the impedance $\mathbf{Z}_-$ of submerged part, generalized velocities $\mathbf{v}_-$ and forces $\mathbf{f}_-$ at the boundary $z = H$ acting from the submerged part of the shell $z = H - 0$. These impedances $\mathbf{Z}_\pm$ connect corresponding vectors of generalized velocities and forces: $\mathbf{Z}_-\mathbf{v}_- = \mathbf{f}_-$ at $z = H - 0$ and $\mathbf{Z}_+\mathbf{v}_+ = \mathbf{f}_+$ at $z = H + 0$. The continuity condition of generalized velocities $\mathbf{v}$ and forces $\mathbf{f}$ on the boundary $z = H$ dividing protruding and submerged parts of the shell has the view: $\mathbf{v}_- = \mathbf{v}_+ \equiv \mathbf{v}, \mathbf{f}_- = \mathbf{f}_+ \equiv \mathbf{f}$. Hence $(\mathbf{Z}_- - \mathbf{Z}_+)\mathbf{v} = \mathbf{0}$. The existing condition of a nontrivial solution of this system has the form $\det(\mathbf{Z}_- - \mathbf{Z}_+) = 0$. Solving this equation one can obtain eigen frequencies and eigen vectors of the system liquid-cylinder.

Meanwhile $\mathbf{v}$ is the value of generalized velocities on the boundary $z = H$. Taking into account the expressions

$$
\mathbf{v} = \left.\left(\mathbf{U}^\partial \mathbf{u}_-\right)\right|_{z=H-0} = \left.\left(\mathbf{U}^\partial \mathbf{U}_-\mathbf{t}_-\right)\right|_{z=H-0} = \left.\left(\mathbf{U}^\partial \mathbf{U}_-\right)\right|_{z=H-0}\mathbf{t}_- \equiv \mathbf{U}_-\mathbf{t}_-
$$

the values of vectors $\mathbf{t}_- = \mathbf{U}_-^{-1}\mathbf{v}$, $\mathbf{t}_+ = \mathbf{U}_+^{-1}\mathbf{v}$ can be obtained for submerged and protruding parts of the shell and consequently therefore one can get the decisive value of eigen modes of submerged and protruding parts of the shell:

$\mathbf{u}_- = \tilde{\mathbf{U}}_-\mathbf{t}_- = \tilde{\mathbf{U}}_-\mathbf{U}_-^{-1}\mathbf{v}$, $\mathbf{u}_+ = \tilde{\mathbf{U}}_+\mathbf{t}_+ = \tilde{\mathbf{U}}_+\mathbf{U}_+^{-1}\mathbf{v}$. Therefore we obtain a completed view of a problem solution in the terms of a displacement vector of the shell $\mathbf{u}(\varphi, z)$ and pressure in the liquid $P(r, \varphi, z)$.

The important point to be made is the particular case of impedances when the protruding part of the shell $H < z \le l$ is absent. Then in the case of free edge at $z = H$ the dispersion equation transform is $\det \mathbf{Z}_- = 0$. Taking into account that the dispersion equation can be rewritten in the terms of admittance matrixes $\det(\mathbf{Z}_+^{-1} - \mathbf{Z}_-^{-1}) = 0$, the following variant of dispersion equation $\det \mathbf{Z}_-^{-1} = 0$ can be obtained in the case of rigidly fixing of the upper edge, $z = H$.

Propagation of a Wave on a Cylinder, Partially Submerged into a Liquid

We consider the case when the protruding part of the cylinder extends to $+\infty$. Let the wave $t_5\mathbf{u}^5$ propagate from $+\infty$ (here t_5 is given amplitude). Then the field of displacements in the protruding part of the shell has the view $\mathbf{u}_+ =$

$\sum_{j=1}^{4} t_j\mathbf{u}^j + t_5\mathbf{u}^5 = \tilde{\mathbf{U}}_+\mathbf{t}_+ + t_5\mathbf{u}^5$ (we must take into account that all waves in the protruding part of the shell have to be chosen according to the principle of limited absorption [17], then, particularly, $t_j = 0$, $j = 6, 7, 8$). The vectors of generalized displacements and forces in this case have the view

$$\left\{ \begin{matrix} \mathbf{v}_+ \\ \mathbf{f}_+ \end{matrix} \right\} = \left\{ \begin{matrix} \mathbf{U}^\partial \\ \mathbf{F}^\partial \end{matrix} \right\} \mathbf{u}_+ = \left\{ \begin{matrix} \mathbf{U}^\partial \\ \mathbf{F}^\partial \end{matrix} \right\} (\tilde{\mathbf{U}}_+\mathbf{t}_+ + t_5\mathbf{u}^5) = \left\{ \begin{matrix} \mathbf{U}_+ \\ \mathbf{F}_+ \end{matrix} \right\} \mathbf{t}_+ + t_5 \left\{ \begin{matrix} \mathbf{v}^5 \\ \mathbf{f}^5 \end{matrix} \right\}.$$

Let us take into account the continuity condition of generalized velocities and forces at $z = H$ and the fact that impedance connects vectors of generalized velocities and forces ($\mathbf{Z}_-\mathbf{v}_- = \mathbf{f}_-$) for the submerged part of the shell at $z = H - 0$. Hence the vector of unknown constants $\mathbf{t}_+$ is determined to be

$$\mathbf{Z}_-(\mathbf{U}_+\mathbf{t}_+ + t_5\mathbf{v}^5) = \mathbf{F}_+\mathbf{t}_+ + t_5\mathbf{f}^5 \quad \Rightarrow \quad \mathbf{t}_+ = t_5(\mathbf{Z}_-\mathbf{U}_+ - \mathbf{F}_+)^{-1}(\mathbf{f}^5 - \mathbf{Z}_-\mathbf{v}^5)$$

and, consequently, the field of displacements in the protruding part of the shell is found. In order to find the field in the submerged part of the shell one can follow the way described in the previous section.

Conclusions

The exact analytical solution of stationary vibrations of a system of coinciding cylinders partially submerged into a liquid is obtained. The solution is written in the form of input impedances. It allows us to consider as the protruding part not only thin shells but even more complicated constructions on the assumption that their input impedances are known. On the basis of obtained solutions one can construct solutions of forced vibration problems as an expansion on appropriate modes of homogeneous problems.

Acknowledgements The author expresses sincere gratitude to Professor Daniil P. Kouzov for useful discussions and constructive notes.

References

1. Zinovieva TV (2012) Computational mechanics of elastic shells'of revolution in mechanical engineering calculations. In: Modern engineering: science and education. Proceedings of second international scientific and practical conference. Published by State Polytechnic University, SPb, pp 335–343
2. Filippenko GV (2012) The vibrations of pipelines and thin walled supports of hydro technical constructions partially submerged into the liquid. In: Modern engineering: science and education. Proceedings of second international scientific and practical conference. Published by State Polytechnic University, SPb, pp 769–778
3. Lavrov JA (1997) On the free vibrations of acoustical resonator with elastic cylindrical wall and rigid flat end. Acoust J 3:425–428
4. Lavrov JA (2002) On the frequencies of free vibrations in the liquid, separating rigid and elastic walls of the finite length. Papers of scientific seminars of PDMI, 2002
5. Filippenko GV, Kouzov DP (2001) On the vibration of membrane partially protruding above the surface of a liquid. J Comput Acoust (JCA) 9(4):1599–1609
6. Filippenko GV (2006) Vibration of the cylindrical supports and tubes partially submerged into the water. In: Proceedings of the 8th international symposium "Transport Noise and Vibration", St.-Petersburg, Russia, 4–6 June 2006, CD format, 2006. Article s2–5, 7 pp. Published by East-European Acoustical Association, ISBN 5-900703-92-4
7. Filippenko GV (2006) Vibration of the cylindrical supports and tubes partially submerged into the water. In: Proceedings of the 8th international symposium "Transport Noise and Vibration", St.-Petersburg, Russia, 4–6 June 2006, CD format, 2006. Article s2–5, 7 pp. Published by East-European Acoustical Association, ISBN 5-900703-92-4
8. Kouzov DP, Filippenko GV (2010) Vibrations of an elastic plate partially immersed in a liquid. J Appl Math Mech 74(5):617–626
9. Yeliseyev VV, Zinovieva TV (2012) Nonlinear-elastic strain of underwater pipeline in laying process. Vycisl. meh. splos. sred.—Comput Continuum Mech 5(1):70–78
10. Filippenko GV (2013) Energy aspects of waves propagation in the infinite cylindrical shell fully submerged into the liquid. Vycisl. meh. splos. sred.—Computat Continuum Mech 6 (2):187–197
11. Sorokin SV, Nielsen JB, Olhoff N (2004) Green's matrix and the boundary integral equations method for analysis of vibrations and energy flows in cylindrical shells with and without internal fluid loading. J Sound Vib 271(3–5):815–847

12. Filippenko GV (2010) Statement of the boundary-contact problems for the shells in acoustics. In: Proceedings of the international conference on "Days on Diffraction 2010", St.-Petersburg, Russia, 8–10 June 2010, pp 57–626
13. Filippenko GV (2011) The energy analysis of shell-fluid interaction. In: Proceedings of the international conference on "Days on Diffraction 2011", St.-Petersburg, Russia, 30 May–3 June 2011, pp 63–66
14. Filippenko GV (2012) The forced oscillations of the cylindrical shell partially submerged into a layer of liquid. In: Proceedings of the international conference on "Days on Diffraction 2012", St.-Petersburg, Russia, 28 May–1 June 2012, pp 70–75
15. Cremer L, Heckl M, Petersson BAT (2005) Structure-borne sound. Structural vibrations and sound radiation at audio frequencies, 3rd edn. XII, 607 p 215 illus
16. Eliseev VV (2003) Mechanics of elastic bodies. SPbSPU, SPb, 336 p (in Russian)
17. Sveshnikov AV (1951) Absorption principle for a wave guide. Doklady Akademiji Nauk SSSR 80:345–347 (in Russian)

Mathematical Modelling of Interaction of the Biped Dinamic Walking Robot with the Ground

Anastasia Borina and Valerii Tereshin

Abstract Walking robots are the most wanted, important and interesting ones and their stability is the most crucial problem these days. Technical progress involved using modern materials and technologies in creating of new walking devices and its control system (Brogliato in IEEE Trans Autom Control 48(6):918–935, 2003 [1]), which could provide static and dynamic stability. Dynamic walking means to constantly fall, but to bring forward the swing leg in time to prevent tilting over (Automatica 35(3):374–535 [2], Collins et al. in Int J Robot Res 20(7):607–615, 2001 [3]). The control system must provide stability of walking and its efficiency. This paper proposes the method of control of biped walking dynamic robot and simulated parameters of walking. The parameters of control are time and place of putting the leg on the ground at the beginning of the next step. For its defining the equations of the spatial turned mathematical pendulum are used.

Keywords Biped walking · Walking robots · Dynamic walking · Control system

Introduction

The walking robots as opposed wheel have following advantages:

– high passableness;
– high adaptation to roughness of a road;
– high mobility;
– comfortable transportation of the person or sensitive equipment.

A. Borina (✉) · V. Tereshin
Peter the Great Saint-Petersburg Polytechnic University,
Saint Petersburg, Russia
e-mail: kamchatka1@rambler.ru

V. Tereshin
e-mail: terva@mail.ru

© Springer International Publishing Switzerland 2016
A. Evgrafov (ed.), *Advances in Mechanical Engineering*,
Lecture Notes in Mechanical Engineering, DOI 10.1007/978-3-319-29579-4_13

127

These properties of walking robots define their practical application [4, 5]. Today it is actual and perspective to use them for transportation of cargoes in cross-country, for search and rescue operations [6], for liquidation of consequences of technogenic and natural accidents, when chemical or radioactive pollution, for working in fires, under water, in the military purposes, for protection, cleaning of houses, for any researches or supervision, for example, on a surfaces of planets of solar system or space bodies. Also, it is necessary to mention about using exoskeleton for returning of impellent activity to people or for increase human efforts [7, 8].

Today many modern interesting walking devices are created. The main problem in this field is control [9, 10]. There are two main types of bipedal walking are present in the literature: static and dynamic. The static walking is always stable, but such robots have low speed and big weight. Dynamic walking is considered to be human like [11, 12]. As opposed static, dynamic robots are faster, more maneuverable and dexterous. The feet of such robot are very small, like dots, so it is impossible to be stable for them.

Equations of Movement

Figure 1 shows the structure of the biped and coordinates used to describe the configuration of the system.

Define geometrical parameters and corners of left foot orientation:

$$f_i(\alpha, \beta, \gamma, x, y, z, x_1, y_1, z_1, \theta_1, \phi_1, k_1, \alpha_1, \beta_1, \gamma_1) = 0, \quad i = 1 \ldots 6, \qquad (1)$$

where θ_1, φ_1—angles of rotation of the left joint; k_1—length of the left leg; α, β, γ—the finite rotation angles of the robot body; x, y, z—coordinates of center of gravity;

Fig. 1 Kinematic model of a biped robot

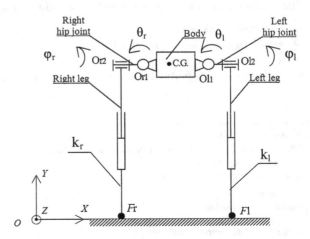

α_1, β_1, γ_1—the finite rotation angles of the left foot; x_1, y_1, z_1—coordinates of pole F_1. For the right foot the equations are similar.

For presenting system of six Eq. (1) in the implicit form we can use method 4×4 matrixes of transition [13]. The system (1) has nine generalised coordinates, let α, β, γ, x, y, z, x_1, y_1, z_1 are independent. So from the system (1) we can define variations of θ_1, φ_1, k_1, α_1, β_1, γ_1. For writing the equations of movement let use a principle of possible moving. To describe of inertial properties of the device it is enough to have weight and a matrix of inertia of the robot body. Describe interaction of foot with a ground only by a force vector. Let consider that the robot leans against the left foot.

$$\Phi_x \delta x + \left(\Phi_y - mg\right)\delta y + \Phi_z \delta z + M^{\Phi}_{zk1}\delta\alpha + M^{\Phi}_{zk2}\delta\beta + M^{\Phi}_{zk3}\delta\gamma$$
$$+ M_{\theta1}\delta\theta_1 + M_{\phi1}\delta\phi_1 + R_1\delta k_1 + F_{x1}\delta x_1 + F_{y1}\delta y_1 + F_{z1}\delta z_1 = 0 \qquad (2)$$

where Φ_x, Φ_y, Φ_z—the main vector of inertial forces projections to axes of absolute coordinate system; M^{Φ}_{zk1}, M^{Φ}_{zk2}, M^{Φ}_{zk3}—the main moment of inertial forces projections to axes of intermediate coordinate systems; mg—robot weight; R_1, $M_{\theta1}$, $M_{\phi1}$—support reaction and the moments in drives; F_{x1}, F_{y1}, F_{z1}—projections of the force vector, operating on foot from the ground, to axes of absolute coordinate system.

Control System

To control the robot walking it is necessary to set the time t_p and to define the place $(x_1; z_1)$ for touchdown at the end of the step and at the beginning of the next. Let use the equations of the turned pendulum:

$$\begin{cases} \dot{x}_{tp} = \frac{k}{2}\left((x_0 - x_1) + \frac{\dot{x}_0}{k}\right)\psi - \frac{k}{2}\left((x_0 - x_1) - \frac{\dot{x}_0}{k}\right)\psi^{-1} \\ \dot{z}_{tp} = \frac{k}{2}\left((z_0 - z_1) + \frac{\dot{z}_0}{k}\right)\psi - \frac{k}{2}\left((z_0 - z_1) - \frac{\dot{z}_0}{k}\right)\psi^{-1} \end{cases} \qquad (3)$$

where $\psi = e^{kt_p}$, $k = \sqrt{g/l}$—frequency of free fluctuations of a mathematical pendulum, l—height of the gravity centre; g—acceleration of free falling; $(x_0, z_0, \dot{x}_0, \dot{z}_0)$—entry conditions; $\dot{x}_{tp}$, $\dot{z}_{tp}$—the set speed in the end of a step.

To define the place of putting leg F it is necessary to work out the nonlinear differential equation (2), the initial conditions are equal to final conditions from the previous step [14, 15]. Figure 2 presents the results of the numerical solutions of the systems of Eqs. (2) and (3) for the case of rectilinear walking with left rotation by the angle 30° and dynamic standing (pacing about a motionless point [16]).

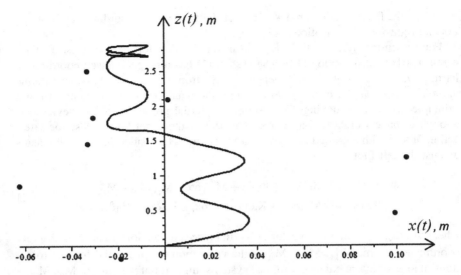

Fig. 2 Trajectory of the centre of gravity during rectilinear walking with left rotation by the angle 30° and dynamic standing

Support Reaction and the Corner of Pressure

After modeling a control system, providing stable walking along various trajectories and dynamic standing it is necessary to set a value of a corner of pressure ϕ_R (Fig. 3).

Figure 4 presents the results of the numerical solutions support reaction R of walking robot during dynamic standing when rate of the step t = 0.37 s, speed in the end of a current step $V_{t1} = 0.46$ m/s.

Fig. 3 Walking device in a one-foot support phase

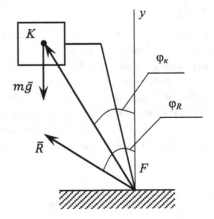

Fig. 4 Temporary dependence of the dynamic standing

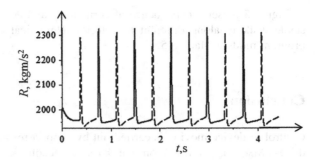

To define ϕ_R, let us use the theorem of movement of the centre of weights of system [17]

$$m\bar{w}_K = \bar{R} + m\bar{g}, \qquad (4)$$

where m—weight of the robot; $\bar{g}$—the acceleration of free fall; $\bar{R}$—support reaction; $\bar{w}_K$—acceleration of the centre of gravity K. The expressions for projections of R to axes of coordinates:

$$R_x = m\ddot{x}, \quad R_y = m(\ddot{y} + g), \quad R_z = m\ddot{z}. \qquad (5)$$

Define ϕ_R from Fig. 3:

$$\phi_R = \text{arctg}\left(\frac{\sqrt{\ddot{x}^2 + \ddot{z}^2}}{\ddot{y} + g}\right). \qquad (6)$$

Define the corner ϕ_K between normal and line FK from Fig. 3:

$$\phi_K = \text{arctg}\left(\frac{\sqrt{x^2 + z^2}}{y}\right). \qquad (7)$$

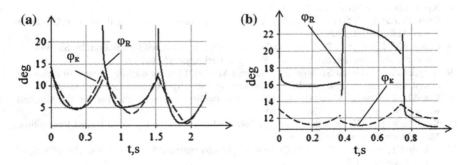

Fig. 5 Temporary dependence of **a** even horizontal walking **b** the dynamic standing

Figure 5 presents dependence of corners ϕ_R and ϕ_K as the results of modeling stable walking along different trajectories. Numerical calculation is executed in program module Maple 15.

Conclusion

Control of device moving is carried out by support reaction R and the moments in drives $M_{\theta l}$, $M_{\phi l}$, formed on the basis of feedback and program movement. Equation (1) allow to define coordinates under indications of gauges of absolute position and orientation of the body and feet.

Length of a foot (KF) and two corners of orientation of a body are stabilised by three drives operated by Pd–regulators [13]. This stabilisation provides small values of vertical fluctuations of a body and turns round perpendicular KF axes. Stabilization of rotation round a vertical is carried out when both feet are on the ground.

At the beginning of the step ϕ_R can be much more a corner of a friction (Fig. 5). Numerical values of these local maxima can be improbable because of imperfection of physical model.

References

1. Brogliato B (2003) Some perspectives on the analysis and control of complementarity-systems. IEEE Trans Autom Control 48(6):918–935
2. Automatica 35(3):374–535
3. Collins S, Wisse M, Ruina A (2001) A 3-d passive dynamic walking robot with two legs and knees. Int J Robot Res 20(7):607–615
4. Brogliato B, ten Dam AA, Paoli L, Genot F, Abadie M (2002) Numerical simulation of finite dimensional multibody nonsmooth mechanical systems. ASME App Mech Rev 55:107–150
5. Lum HK, Zribi M, Soh YC (1999) Planning and control of a biped robot. Int J Eng Sci 37:1319–1349 (Mabrouk, M)
6. Chevallereau C, Aoustin Y (2001) Optimal reference trajectories for walking and running of a biped robot. Robotica 19:557–569
7. Chevallereau C (2003) Time-scaling control of an underactuated biped robot. IEEE Trans Robot Autom 19(2):362–368
8. Gienger M, LNoOer K, Pfei:er F (2003) Practical aspects of biped locomotion. In: Siciliono B, Dario P (eds) Experimental robotics. Springer, Berlin, pp 95–104
9. Brogliato B, Niculescu SI, Monteiro-Marques M (2000) On the tracking control of a class of complementarity–slackness hybrid mechanical systems. Syst Control Lett 39:255–266
10. Das SL, Chatterjee A (2002) An alternative stability analysis technique for the simplest walker. Nonlinear Dyn 28(3):273–284
11. Kar DC, Kurien IK, Jayarajan K (2003) Gaits and energetics in terrestrial legged locomotion. Mech Mach Theory 38:355–366
12. Saidouni T, Bessonnet G (2003) Generating globally optimised sagittal gait cycles of a biped robot. Robotica 21:199–210

13. Borina AP, Tereshin VA (2012) The spatial movement of the walking device under the action of the control actions from the side of the legs. In: 2nd international scientific and practical conference on "modern engineering: science and education". Publishing House of the Polytechnic University, St.-Petersburg, pp 177–178 (in Russian)
14. Borina A, Tereshin V (2015) Control of biped walking robot using equations. Advances in mechanical engineering, Lecture notes in mechanical engineering. Springer, Switzerland, pp 23–31
15. Borina AP, Tereshin VA (2014) Rectilinear uniform movement of the humanoid robot. XL international scientifically-practical conference "Science Week of SPbSPU" (Part 1). Publishing House of the Polytechnic University, St.-Petersburg, pp 160–162 (in Russian)
16. Borina AP, Tereshin VA (2013) Solution of the problem of spatial motion of a statically unstable walking machine. In: 3rd international scientific and practical conference on modern engineering: science and education. Publishing House of the Polytechnic University, St.-Petersburg, pp 631–641 (in Russian)
17. Dankowicz H, Adolfsson J, Nordmark A (2001) 3D passive walkers: finding periodic gaits in the presence of discontinuities. Nonlinear Dyn 24:205–229

Programmable Movement Synthesis for the Mobile Robot with the Orthogonal Walking Drivers

Victor Zhoga, Vladimir Skakunov, Ilya Shamanov
and Andrey Gavrilov

Abstract A mathematical model was developed to support decisions for choosing an algorithm of robot motion, depending on the system of the spatial orientation's data. Programmable continuous movement of a robot is considered in detail in each of its phases. The developed software for synthesis of robot's movement is considered. The simulation results are used for the synthesis of system signals that control the motion, parametric optimization of its geometric, mass and energy characteristics.

Keywords Robot with orthogonal drivers · Mathematical model · Algorithms of robot motion · Control system

Introduction

Mobile autonomous robots, equipped with appropriate hardware, have been designed to perform tasks in hard-to-reach places and in environments that are life-threatening for humans [1]. To perform assigned tasks in a non-deterministic space, a mobile robot needs to first form some general tasks, and then conduct a local analysis of the terrain. Information about the environment enables spatial orientation, movement, and the required technological manipulations. To this end, the robot is equipped with a perception sensor system for operations of general and local environmental analysis. An information-measuring system includes sensors, a

V. Zhoga (✉) · V. Skakunov · I. Shamanov · A. Gavrilov
Volgograd State Technical University, Volgograd, Russia
e-mail: zhoga@vstu.ru

V. Skakunov
e-mail: svn@vstu.ru

I. Shamanov
e-mail: iv.shamanow@gmail.com

A. Gavrilov
e-mail: krobotech@gmail.com

© Springer International Publishing Switzerland 2016
A. Evgrafov (ed.), *Advances in Mechanical Engineering*,
Lecture Notes in Mechanical Engineering, DOI 10.1007/978-3-319-29579-4_14

135

vision system and an on-board processor and is used to implement the algorithms of decision making in autonomous and supervisor management modes with elements of artificial intelligence. In complex environments, an opportunity is provided for remote robot control by the operator [2].

To solve navigation problems, the robot is equipped with an infrared and ultrasonic range finder and an infrared camera of structured light. The system of decision-making support and robot motion control must be able to perform the following operations on the basis of the received information:

- carry out pre-processing of the collected information under continuous control of the robot movement using its own information—measuring system;
- to solve problems of local and global navigation and managing state of the robot according to the sensory system data;
- to perform the necessary conversions of data in information-measuring system for routing robot movement in supervisory management or autonomous mode.

The models of electromechanical multilink systems of a robot exist in an external environment. Data required to solve specific problems and algorithms for their implementation are loaded in the computer beforehand. During the robot movement, a computing device receives information about the current state of the electromechanical system of the robot and information about the dynamic state of the environment. On the basis of this information and its own software resources, control commands for the executing units of the robot are formed to implement the solution of the tasks set.

The mathematical model of the dynamics of a spatial robot movement, taking into account the real properties of the external environment and the design parameters of which are calculated in real time, occupies one of the central places in the decision support for implementation of the task [3, 4].

The Mechanical Design of the Robot

An autonomous mobile robot (Fig. 1) consists of a top body connected to the rotation mechanism with a bottom body [1]. The power unit and control system unit are placed on the robot body. The power of the robot is supported from batteries. On each of the body parts guide sleeves are fixed (Fig. 1). In these sleeves movable frames with the possibility of horizontal movement are mounted associated with the drivers of the horizontal movement. At the ends of the frame, retractable vertical support columns are mounted, interacting with the supporting surface.

An autonomous mobile robot has the following technical characteristics: weight of the robot is 184 kg; maximum load capacity of the robot—2000 N; overall dimensions $1.29 \times 0.764 \times 0.857$ m; type of propulsion—walking orthogonal; the maximum speed is 1.5 km/h; drive type—electromechanical; overcoming the obstacles: width of the ditch—0.55 m, height of threshold is 0.4 m, the angle of elevation—$22°$, (while maintaining the horizontality of the body), stairs

Fig. 1 Robot movements
with orthogonal drivers:
1 upper body, *2* lower body,
3 guide sleeves, *4* driver of
the horizontal movement,
5 robot frames with support
columns

(step height, m/inclination, deg.)—0.4/22; radius of action: when operating wire-lessly up to 500 m power supply—batteries 2×12 V, 55 A/h, the duration of work—2.5 h, turn in place for 360°.

The advantages of the chosen design consist in the combination of metal structures and drivers, resulting in improved specific capacity and energy of mechanisms. Structure and functions of the control unit are determined by the nature of program motions of the considered walking robot.

Programmable Movements

Elementary displacements of a robot can be divided into two groups: course movement and maneuvering. In a course mode of motion, the kinematic scheme of the robot with orthogonal drivers allows one to move the robot body in start-stop driving modes (Fig. 2). These gaits are characterized by the presence of intervals during which no relative movement of the robot frames take place. The necessity of these phases is due to the requirement of unstressed contact of the vertical columns when adapting the robot to the bearing surface.

All straight movement gaits of the robot are periodic. Rectilinear movement of the robot on a terrain that does not contain obstacles comparable with the linear dimensions of the robot is as follows. In the initial position (I) the robot is at rest and is supported by four support columns (Fig. 2). Supporting columns of the other frame are raised to a half stroke.

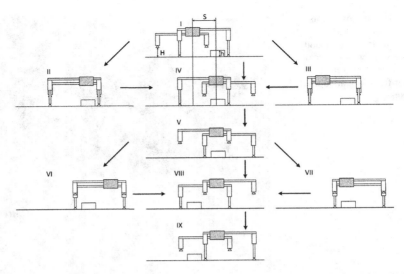

Fig. 2 Example of implementation of a start-stop moving algorithm with a step equal to S

Phase 1. With the help of actuators of horizontal displacement, the body is displaced by the distance S and a no-base frame 2 is displaced by the distance S relative to the body. The end of this phase corresponds to the state II of the robot.

Phase 2. The non-supporting frame columns are lowered until touching the supporting surface, and the columns of the other frame are raised up to the middle position (condition III), while a horizontal leveling system provides the horizontal position of the frames of the robot.

Phase 3. With the help of actuators of horizontal displacement, the body with frame is displaced by the distance S and the frame is displaced by the distance S relative to the body. (state IV of the robot).

Phase 4. The non-supporting frame columns are lowered until they touch the supporting surface, and the columns of the other frame are raised up to the middle position, while the horizontal leveling system provides a horizontal position of the robot's frames. This (condition V) corresponds to the initial position of the robot (state I). At then end of this cycle, movement ends.

Versions of gaits with movement of the body at the same distance 2 s per cycle, with alternately movement of the body and robot frames (the sequence of states $I \rightarrow VI \rightarrow II \rightarrow III \rightarrow VIII \rightarrow IV \rightarrow V$, $I \rightarrow VI \rightarrow II \rightarrow III \rightarrow IX \rightarrow IV \rightarrow V$, $I \rightarrow VII \rightarrow II \rightarrow III \rightarrow IV \rightarrow V$, $I \rightarrow VII \rightarrow II \rightarrow III \rightarrow VIII \rightarrow IV \rightarrow V$, $I \rightarrow VII \rightarrow II \rightarrow III \rightarrow IX \rightarrow IV \rightarrow V$, $I \rightarrow VI \rightarrow II \rightarrow III \rightarrow IV \rightarrow V$ and symmetrical for them), which can be demanded depending on the parameters of the local obstacles of support surface and the nature of the tasks being solved shown on Fig. 2.

The kinematic scheme of the robot also allows one to implement a gait in which the robot body moves continuously, and unstressed adaptation of the robot to the irregularities of the support surface is provided by programmable laws of relative movement of its parts [5].

A kinematic scheme of a robot allows maneuvering in several ways: discrete rotation at which the center of mass of the robot does not move; discrete rotation at which the center of mass is moved by a broken curve; the forward movement of one frame relative to another in straight-line movement of the body. The characteristics of the relative movements of the parts of the robot in discrete ways of turning and during the forward movement are independent and determined by the specific conditions of operation of the robot. In particular, the spatial position of the robot, the external environment state, the assessment of which is carried out according to the vision system data, are taken into account.

The robot movement in the mode in which the speed of the body in the coordinate system associated with a supporting surface remains constant, occurs according to the following algorithm.

Before the movement control system ensures the relative position of the elements of its construction, in which the body of the robot is distanced from the extreme position relative to the supporting frame in the direction opposite to the direction of movement on the magnitude Δ and the frame that does not have at the moment any contact with the supporting surface, is in its extreme position in the direction reverse to the direction of motion (Fig. 3).

Phase 1. From the initial position (Fig. 3) the non-supporting frame 1 moves with speed V_1. The time at which frame 1 will go from one of its extreme positions to the other, i.e. moves a distance S relative to body 3, which in relative motion must move by a distance $[S - 2\Delta]$ as shown in Fig. 4. That is, the body 3 is in this phase moving relative to the frame 1 at an average speed $V_2 = \left[\frac{S-2\Delta}{S}\right] V_1$, this speed is provided by the drive of horizontal movement of the support frame.

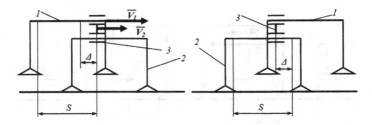

Fig. 3 The mode of the continuous movement of the body, phase 1

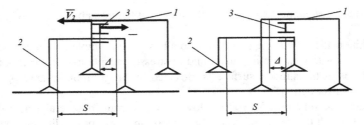

Fig. 4 The mode of the continuous movement of the body, phase 2

Phase 2. In the second phase of the movement (Fig. 4), in order to change the supporting struts, you need to ensure that the absolute speed of free frame is zero. The body continues to move at a speed V_2. To this end, when the free frame reaches its extreme position, the drive of the reverse longitudinal movement of the frame 1 turns on and the movement is made in the opposite direction at a speed V_2. While the body is passing the Δ distance, the struts of a non-supporting frame 1 are lowered until they touch the supporting surface, yet the horizontal position of the robot is ensured, and the struts of frame 2 are raised to the middle position. The design of the robot allows movement of the body along the guides of the frames, even if all the supports of both frames are in contact with the supporting surface.

Phase 3. During this phase (Fig. 5) non-supporting frame 2 moves with speed V_1 and goes from its one extreme position to the other, i.e. moves at a distance S relative to body 3, then body 3, in relative motion moves at a certain distance $[S - 2\Delta]$. Body 3 in this phase of the motion moves at an average speed of V_2 relative to the frame 2.

Phase 4. In the fourth phase of the movement (Fig. 6), as well as in phase 2, to change the supporting struts, the equality to zero of the absolute velocity of the free frame is ensured. To ensure this condition, a constant equality of speeds of the body and the free frame by modulo, and their opposites on a sign in the same coordinate system are needed (similar to phase 2). The supports of non-supporting frame 2 are lowered until they touch the

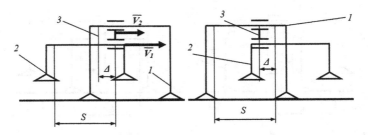

Fig. 5 The mode of the continuous movement of the body, phase 3

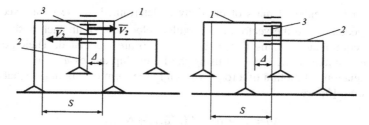

Fig. 6 The mode of the continuous movement of the body, phase 4

supporting surface, and the supports of frame 1 are raised to the middle position, this corresponds to the initial relative position of the parts of the robot. The horizontal leveling system provides the horizontal position of the frame of the robot.

Mathematical Model of the Motion Dynamics

A mobile robot with walking propulsors is a complex multi-link electromechanical system. The mechanical part is represented in the form of six rigid bodies with masses $m_1, m_2, m_3, m_4, m_5, m_6$ associated with the executive DC drives. The movement of the robot relative to a fixed reference system $O\xi\eta\varsigma$ (Fig. 7) is considered. The coordinate axes C_1XYZ are associated with the center of mass of the bottom body and move progressively relative to the fixed coordinate system. The

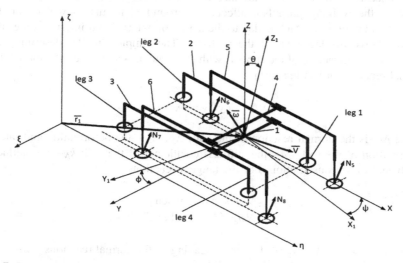

Fig. 7 Design scheme of a mobile robot

position of the center of mass of the body is determined by the radius vector $\vec{r}_1$, velocity vector $\vec{V}$ and the vector of its angular velocity $\vec{\omega}$. With each of the solids, movable coordinate systems $C_k X_k Y_k Z_k$ $k = 1 \div 6$ are associated. Their positions relative to the axes $C_1 XYZ$ are set using the Euler ship angles φ, ψ, θ.

The equations of motion of a robot written in the form of Lagrange equations of the 2nd kind [6] are

$$A\ddot{q} = B(q, \dot{q}) + F(q, \dot{q}, t) + N(q, \dot{q}) \tag{1}$$

where A—$n \times n$ the symmetric matrix of inertia coefficients; $B(q, \dot{q})$—matrix-column of components that depend on the generalized coordinates and velocities; $F(q, \dot{q}, t)$—matrix-column of control forces, forces of gravity and resistance; $N(q, \dot{q})$—matrix-column of reaction forces of support surface and moments of these forces about axes $C_1 XYZ$, the matrix dimension depends on the type of program motion of the robot.

Base plates of the columns are in contact with the bearing surface in points and the reaction of the surface is reduced to the forces $\vec{N}_i$, $i = 1 \div 8$.

To determine the forces of reactions of the base plate interaction with the ground, absolute coordinates ξ_i^K, η_i^K, ς_i^K of the entry point of its contact with the supporting surface are determined, i.e. the coordinates of that point that simultaneously satisfies equations of the trajectory of a reference point and ground surface.

To fulfill the conditions of stability of the integration, the moment when the reference point touches the ground is determined at the point of intersection of the line $[(\varsigma_i^K)^n, (\varsigma_i^K)^{n+1}]$ in n integration steps with the equation of the surface $f(\xi_i^K, \eta_i^K)$. Upon further interaction of the support column with the ground, the projection of the reaction force on the axis of the fixed coordinate system $O\xi\eta\varsigma$ is determined in accordance with the adopted rheological soil model.

When the walking propulsor affects the ground, deformation will occur, the amount and nature of which is due to the action of external and internal forces that cause relative displacement of the particles. The simplest model describing the nature of the interaction of the support with the ground, is a visco-elastic model. For normal reactions it has the form

$$N = c_N \Delta_N + k_N \dot{\Delta}_N, \tag{2}$$

where N—is the normal reaction of the soil; Δ_N, $\dot{\Delta}_N$—the deformation and rate of deformation of the soil; c_N—the stiffness coefficient of the soil; k_N—the adduced coefficient of viscous friction. For the tangent reactions

$$T = \begin{cases} c_\tau \Delta_\tau + k_\tau \dot{\Delta}_\tau, & \text{when } |T| < T_p, \\ f_{ck} N sign(\dot{\Delta}_\tau), & \text{when } |T| \geq T_p, \end{cases} \tag{3}$$

where Δ_τ, $\dot{\Delta}_\tau$, c_τ, k_τ – have the same meaning as for normal reactions; T_p—static friction force; f_{ck}—the coefficient of sliding friction.

The adopted soil model is an approximate, but its use is justified not only by simplicity, but also by a weak correlation between the nature of the soil deformation and the majority of the analyzed parameters of robot motion.

Software Synthesis of Robot Movements

To simplify the mathematical description of robot movements, its structural scheme is presented in the form of a system with variable structure.

The unchanged part is the robot body m_1, m_4, and the variable parts are guide sleeves m_2, m_3, m_5, m_6, which may be deemed to be a part of the body, or separate components of the model depending on the phase of movement [7]. This modification of the design scheme allows us to model the physical design of elastic couplings between the guides and frames of the robot. The elastic ties accounting leads to oscillations of high frequency.

The body corresponds to the first 6 equations of the math model, which specify the displacement in the reference system $O\xi\eta\varsigma$ and Euler ship angles φ, ψ, θ and are calculated under any conditions.

The remaining 4 equations model describes the relative movement of the guides sleeves and the robot body and is only considered with the activity of the respective actuators. One equation corresponds to the relative movement of one guide.

The calculation of the robot only on the equations of its body is made in moments of absence of relative movement between the guide and the body, i.e. at a state of rest or during a change of support columns. Since the movement of the guide adds one equation, in the moving phases in start-stop mode 8 equations are integrated: 6 equations of the body and two equations of movement of the guides. In the continuous movement of the body the full system of 10 equations is integrated. This is the costliest mode from the point of view of computing resources.

The calculation algorithm used allowed us to significantly reduce the number of calculations. Besides, this algorithm allowed us to take into account lack of horizontal movement of the body along the guide without the introduction of additional terms and to implement the immobility of the body in the reference system $O\xi\eta\varsigma$ when guides are moving.

The algorithm used allowed us to perform the numerical integration of the differential equations using a simple Runge-Kutta algorithm of 4-th order, since lack of high-frequency components provides sufficient accuracy in integration in real time. For finding the coefficients of the method at each step of the calculation, a system of linear algebraic equations is solved. The solution is found using a Gauss method with allocation of the main element, which reduces the error of computation in comparison with the conventional method of Gauss.

The software is written in C# with the use of the library OxyPlot for output graphs. The program includes the following modules: a GUI operator and a dynamic loaded library that calculates the model. Extraction of calculations into a separate library will allow us to easily integrate this model into a system of autonomous robot control in the future.

The structure of the program is shown on an UML package diagram (Fig. 8).

The GUI package contains classes that implement a graphical operator interface. The Model Solvers package contains classes that implement the method of Runge-Kutta for integration of systems of differential equations and the Gauss

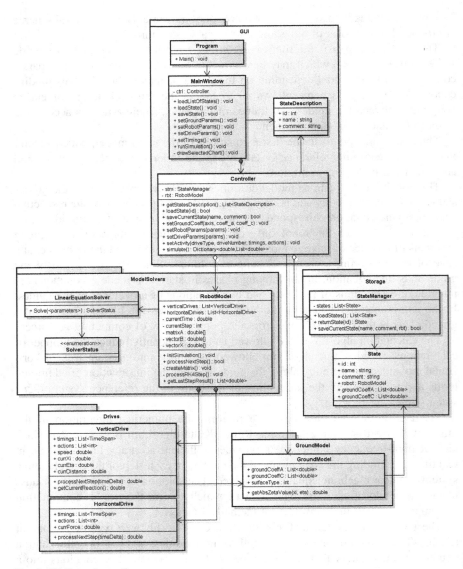

Fig. 8 UML package diagram

method for solving systems of linear algebraic equations, as well as helper functions for the formation of matrices A, $B(q,\dot{q})$, $F(q,\dot{q},t)$ and $N(q,\dot{q})$. Package Drives includes classes for simulating the operation of the actuator horizontal movement and retractable vertical supports, the class in the package Ground Model implements the selected soil model.

The Storage package implements mechanisms of storing and loading of the model's current state, which so carry out complex simulations step by step.

The Results of Software Movements Simulation

The developed software calculates the robot movement on the basis of the math model in one of the modes: start-stop mode or continuous movement of the robot body. In each case, the operator has access to the graphs of load on the drivers of vertical adaptation and tilt angles of the robot relative to the surface. The operator can set the parameters of the soil and the surface topography based on data from the sensory system of the robot, for a better approximation of the simulation to real conditions.

An example of numerical simulation is presented in Fig. 9.

The graph shows the time dependence of the support column reaction for one cycle of movement in start-stop mode while the robot body is in the middle position at the beginning of the movement. On the graph one can highlight 8 characteristic phases present in all start-stop modes of movement. Phase 1—the mode of calculation of the initial conditions. In this phase the initial conditions are automatically determined, the robot does not move. Reactions of support surface are calculated.

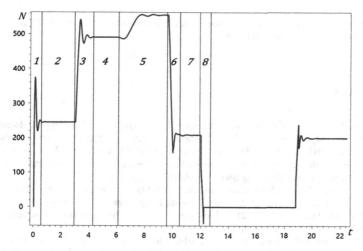

Fig. 9 Graph of the column reaction during one cycle of movement in start-stop mode

A system of six differential equations is integrated. This is a preliminary phase. Phase 2—the robot relies on all 8 columns; the load is evenly distributed among all supports. It's the initial phase for the implementation of the program motion. Phase 3—lift mode of columns of a single frame. In this phase robot detaches from the deformed surface and lift supports of two frames of horizontal movement. A system of six differential equations is integrated. Phase 4—the robot is supported by 4 support columns. A system of six differential equations is integrated. Phase 5— move of non-support modules of horizontal displacement. A system of eight differential equations is integrated. Phase 6—mode of lowering of the support columns. Phase 7—the robot is based on 8 support columns. Phase 8—lift of support columns. While moving the robot in phases 6–8 a system of six differential equations is integrated. This completes the cycle of one-step movement of the robot.

Comparison of simulation results of robot motion, taking into account the real relations between its parts and the proposed method, showed that in order to sustain the integration process, it is necessary to use integration methods for stiff systems. In this case the integration time is increased approximately on the order.

Conclusion

Developed software allows simulation of the behavior of a robot on different types of soil and terrain. Output parameters allow the operator of the robot to make decisions about the selection of a driving mode based on the data of a math model. In general, use of this software will simplify the task of selecting the driving mode and can warn the operator about possible robot failure in an environment, unintended by the design of the mechanisms.

Acknowledgements The work was done with the financial support of RFBR (contract No. 15-01-04577-a, No. 15-08-04166-a).

References

1. Zhoga V, Gavrilov A, Gerasun V, Nesmianov I, Pavlovsky V, Skakunov V, Bogatyrev V, Golubev D, Dyashkin-Titov V, Vorobieva N (2014) Walking mobile robot with manipulator-tripod. In: Proceedings of romansy 2014 XX CISM-IFToMM symposium on theory and practice of robots and manipulators. series: mechanisms and machine science. Springer International Publishing Switzerland, vol 22, pp 463–471
2. Roy SS, Pratihar DK (2013) Dynamic modeling, stability and energy consumption analysis of a realistic six-legged walking robot. Robot Comput Integr Manuf 29(2):400–416
3. Silva MF, MacHado JAT (2012) A literature review on the optimization of legged robots (Review). JVC/J Vib Control 18(12):1753–1767
4. Zhoga VV, Kravchuk A, Skakunov VN (2014) Navigation algorithm for mobile robot using computer vision system with structured-light camera. Innovation Information Technologie:

mater. of the 3rd Int. scien.-pract. conf. (Praque, April 21–25, 2014). Part 2. MIEM VShE, Ros. tsentr nauki i kulturyi v Prage. Moscow, pp 142–144. (In Russian)

5. Zhoga VV, Aniskov RV, Merkulov AA, Skakunov VN (2014) Control system of electric-powered walking robot with orthogonal drivers. Izvestiya VolgGTU. Seriya "Aktualnyie problemyi upravleniya, vyichislitelnoy tehniki i informatiki v tehnicheskih sistemah". Vyip. 21: mezhvuz. sb. nauch. tr.Volgograd, № 12 (139), pp 157–162. (In Russian)

6. Zhoga VV, Skakunov VN, Filimonov AV, Golubev DV (2013) Dynamics of the marching modes of motion of the robot with orthogonal movers. Izvestiya VolgGTU. Seriya "Aktualnyie problemyi upravleniya, vyichislitelnoy tehniki i informatiki v tehnicheskih sistemah". Vyip. 16: mezhvuz. sb. nauch. st. Volgograd, 2013. № 8 (111), pp 14–22. (In Russian)

7. Andreev AE, Zhoga VV, Serov VA, Skakunov VN (2014) The control system of the eight-legged mobile walking robot. Knowledge-based software engineering. In: Proceedings of 11th joint conference, JCKBSE 2014 (Volgograd, Russia, September 17–20, 2014). In: Kravets A, Shcherbakov M, Kultsova M, Iijima T. Volgograd State Technical University [etc.]. Springer International Publishing, pp 383–392. (Series: Communications in Computer and Information Science, vol. 466)

Processing of Data from the Camera of Structured Light for Algorithms of Image Analysis in Control Systems of Mobile Robots

Vladimir Skakunov, Victor Belikov, Victor Zhoga
and Ivan Nesmiynov

Abstract The article discusses the preprocessing algorithms of three-dimensional clouds of points from structured light cameras, methods of image recognition, algorithms for constructing maps and route planning for robot motion. Various methods and algorithms for the transformation of a three-dimensional cloud of points leads to reduction of computational complexity of all stages of data processing for implementation of the algorithm on an onboard navigation system with limited computing resources. Based on the proposed sequence of image conversion, software has been developed for implementation of the algorithm of spatial orientation of a robot. Image processing methods and functions of PCL and ROS metapackage libraries were used. Experimental study of the possibilities of the developed image processing and route construction algorithm was made using the prototype of a mobile robot on a wheeled platform.

Keywords Structured-light camera · 3D point clouds · Pattern recognition algorithms · Supervisory management · Navigation algorithms · Mobile robot

V. Skakunov (✉) · V. Belikov · V. Zhoga
Volgograd State Technical University, Volgograd, Russia
e-mail: svn@vstu.ru

V. Belikov
e-mail: belikov-viktor@mail.ru

V. Zhoga
e-mail: zhoga@vstu.ru

I. Nesmiynov
Volgograd State Agrarian University, Volgograd, Russia
e-mail: ivan_nesmiyanov@mail.ru

© Springer International Publishing Switzerland 2016
A. Evgrafov (ed.), *Advances in Mechanical Engineering*,
Lecture Notes in Mechanical Engineering, DOI 10.1007/978-3-319-29579-4_15

149

Introduction

The practice of image processing and recognition systems application shows that the current level of technology in robotics does not allow us to create fully autonomous systems. The degree of effectiveness of the vision system (VS) is defined by technical characteristics of sensors as well as by methods of information extraction. On the first point we need expensive sensors (time-of-flight cameras based on ToF technology; laser scanners, television cameras, thermo vision cameras and others), and on the second point—high performance computing. Currently within acceptable cost of a vision system for implementation of image processing algorithms and solving problems of spatial orientation, the greatest effect can be obtained in two areas of application of VS: when using the supervisory control of mobile robots and for an autonomous mode of a robot, in which the goals of management are significantly simplified.

Control of the robot in autonomous mode depends on the settings of the sensor cameras: television and time-of-flight cameras, cameras in the infrared range. New opportunities in image processing appear in systems based on structured light cameras, which allow us to obtain three-dimensional pictures of the environment. These include, in particular, a sensor cameras MS Kinect and ASUS Xtion PRO Live, which are based on Prime Sense technology.

Image Recognition Using Structured Light Cameras

The results of software implementation of image processing algorithms based on analysis of the original three-dimensional clouds of points from the structured light cameras showed that, in a number of tasks, procedures for preprocessing and machine learning of classifiers have high priority [1]. To reduce computational complexity, rapidly rising with an increase in the number of points in the processed data, the raw data obtained from the structured light camera should be pre-filtered without significant loss of information about the external environment. In [1] in the image processing algorithm for obstacle identification, the following basic order of operations is proposed: the removal of redundant points, noise reduction, a decrease in the density of the cloud, highlighting the main planes, building a descriptor for the point clouds, classification of objects, and estimation of the distance to the object. The proposed algorithm is implemented in an onboard system for image processing based on the Intel SU7300 processor with a clock frequency of 1.3 GHz and a sensor camera MS Kinect using OpenCV libraries. A series of experiments were performed on recognition of the typical obstacles: arbitrary three-dimensional objects of simple geometric shape, doorway, stairs. During the experiments we measured the time spent on individual steps of the algorithm and the total processing time of one frame, which did not exceed 0.7 s, and the accuracy of recognition which was not less than 0.65. The use of industrial computer ARBOR

FRC-7700 with Intel Core i7 CPU allowed us to improve greatly the vision system work. However, even in this case the possibility of an on-board system built on a powerful single board of computers without graphics accelerators and specialized solvers on FPGA in many practical applications may not provide the work of the image recognition and robot control systems in real-time.

At the same time, studies have shown that onboard systems with much fewer computing resources can autonomously perform various tasks formulated by the remote user, if the additional condition of VS work is within a relatively simple deterministic external environment with a given algorithm of motion and/or known destination. Within these limitations, management tasks during autonomous behavior of a robot can be partially implemented. In this case the problem of identifying obstacles, building maps of the locality and formation of the travel route of the robot to a given target are also significantly simplified.

The VS was built for experimental verification of one of several possible variants of the image processing system, including a sensor controller, ASUS Xtion PRO Live and the microcomputer Raspberry Pi 2. Methods of image processing for this system where chosen out of the limited processing power of the onboard computer.

The operating system Debian Linux was installed on the minicomputer. The open source image processing libraries were used: a library for solving typical robotic tasks ROS Groovy Galapagos and a library for working with point clouds PCL.

Methods of Point Clouds Pre-processing

Processing of images received from structured light cameras when solving navigation tasks are usually divided into three phases: detection of the image, mapping the terrain and route planning of the robot motion in the environment with obstacles.

Given that the on-board system has limited computational resources, it requires prior conversion of the raw data from the sensor cameras to reduce the complexity of their processing in the future. In this paper, three types of conversions of the source data are used: a decrease in density, clipping points and their projection onto a plane.

To illustrate the successive stages of transformation of the image the scene shown in Fig. 1 is used.

The decrease in the density. The original point cloud is very dense, which increases data processing time. To reduce the detail of the point cloud, a three-dimensional voxel grid is constructed. Each group of points within one cell is approximated as the centroid of that group. This method is slightly slower than the approximation of the center of the cell, but allows us to reduce the distortion [2]. As can be seen from the example (Fig. 2), information about the shape of the obstacles is not lost.

(a)   **(b)**

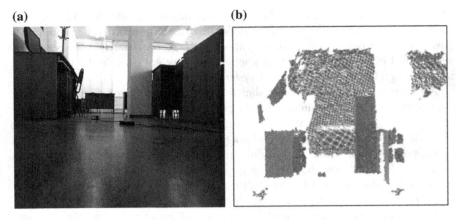

Fig. 1 The camera image (**a**) and its corresponding point cloud (**b**)

(a)   **(b)**

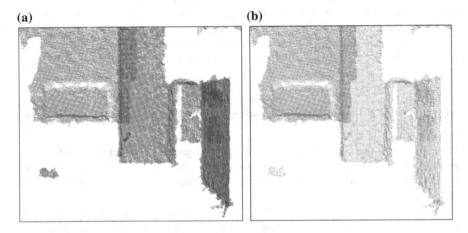

Fig. 2 Point cloud before (**a**) and after (**b**) applying a filter of the density decrease

Clipping points. Some of the points in the cloud are discarded because they do not contain information about potential obstacles. These include points that lie above or below a predetermined elevation, and points lying on a particular distance, for after this value (4 meters or more) measurement error increases, and perhaps the appearance of the "ghost" of obstacles that actually may lie much further. An example of the filter work is shown in Fig. 3.

Projection on the plane. The projection of a cloud on the plane of motion of the robot means a transition to a two-dimensional image that allows one to dramatically reduce the computational complexity of the methods used.

(a)   (b)

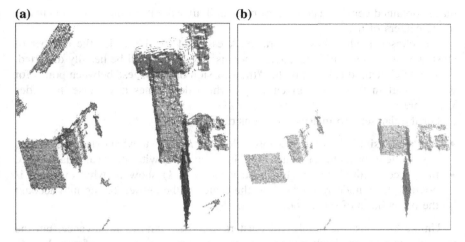

Fig. 3 Clipping points, the source cloud (**a**) and the cloud after the filter (**b**)

An Algorithm for Constructing Maps of the Environment

The proposed algorithm of map constructing consists of several stages:

- clustering of points;
- constructing the minimal convex hulls;
- space representation in a form suitable for navigation.

Features of algorithms appear on separate stages.

Points clustering algorithms. Clustering of points in a cloud is necessary to highlight certain obstacles. In developing the algorithm, three possible methods of clustering were considered: APC method (Affinity Propagation Clusters), k-means, DBSCAN algorithm.

K-means method. The method is based on minimizing the mean square deviation of the cluster points from their centers of mass by the formula [3]

$$V = \sum_{i=1}^{k} \sum_{x_j \in S_i} (x_j - \mu_i)^2$$

where
k number of clusters;
S_i clusters obtained;
μ_i is the center of mass vectors.

At the initial stage the centers of mass are given arbitrarily. At each iteration the centers of mass are calculated afresh for the clusters and all the points are collected on the obtained centers. The process runs until after the iteration shows no change in the centers of mass.

The chosen method has one drawback essential for this task—the number of clusters must be known in advance, otherwise the results will be heavily distorted.

The APC method is based on the "transmission of messages" between points for which a certain function is given $s(i, j)$, which determines how close the nodes x_i и x_j are.

For the dataset two matrices are constructed:

- the "responsibility" matrix R whose values $r(i, k)$ show to what extent the node x_k is suitable to be a cluster center for x_i in comparison with other candidates for x_i
- the "accessibility" matrix A, whose values $a(i, k)$ show to what extent it is possible for a node x_i to take x_k as the center of the cluster, taking into account the preferences of other nodes for x_k.

Unlike k-means, the number of clusters may be unknown in advance, but the method has another disadvantage: when composing two matrices of size $N \times N$, where N is the number of points, it requires a huge size of memory for large data sets. In conditions of limited computational resources of an on-board system and point clouds containing thousands of points, this method is not applicable.

The DBSCAN algorithm allocates clusters based on reachability by dense regions [4]. Point p is directly reachable from a point q if the point p is located not further than a certain distance ε from the point q and the point q is surrounded by a sufficient number of points N_{min}.

The points p and q are considered to be reachable if there exists a sequence of points $p_1, p_2, ..., p_n$, where $p_1 = p$, $p_n = q$, and any points p_i and p_{i+1} are directly reachable from each other.

Since reachability is not symmetric, we introduce the concept of dense connectivity. The points p and q are connected if there is such a point o, reachable both from p and from q. Thus, a cluster is a set of points densely connected with each other.

The calculations performed for the clustering of points in the handling of images that differ in complexity of the scene show the advantages of a DBSCAN algorithm on the following criteria: the number of clusters may be unknown in advance, the requirements for computing resources are low, and the density condition allows the algorithm to work effectively in noisy data sets, which is especially important when processing data from structured light cameras.

Construction of geometric models. Obstacles can be represented geometrically in the form of polygons—minimal convex hulls around clusters of points. For their construction several algorithms are applied.

Discrete partitioning of space. One approach is to divide the whole space into subspaces that are grouped into passable and impassable areas. A busy net is a simple discrete space fragmentation—the division into cells which are considered passable or not. Although this gives a very simple data structure, this method has several drawbacks. This group includes construction of trapezoidal map. Maps of this type are less redundant than the busy net due to the fact that the splitting points are not distributed uniformly throughout the space, but are concentrated in the area of obstacles and other important objects. The routes based on such partitioning are not optimal, as they often depend on the chosen space restrictions of the map.

The Graham Scan. The algorithm is to bypass the stack of points, sorted in increasing of polar angle clockwise relative to the point with the minimum ordinate (and minimum abscissa, if there are several such points) [5].

At each step the condition is checked that, when moving between points, the left turn is not performed. The condition of left turn between the three points a, b, c will look as follows: $u_x v_y - u_y v_x > 0$, where $u = \{b_x - a_x, b_y - a_y\}$, $u = \{c_x - a_x, c_y - a_y\}$. If a turn is performed, then the point is discarded and a step is repeated.

After a full bypass of the stack, only the points that remain constitute the minimum convex hull for a given set of points.

The visibility graph construction. The vertices of this graph are the vertices of the polygons, the edges are the sides of the polygons, and the lines connecting the polygon vertices to other vertices are visible from them.

We can construct a Lee algorithm, which allows us to build a graph in $O(n^2 \log n)$, where n is the number of points. In this algorithm the sweeping straight line is used. The straight line comprises segments that cross the sweeping ray, which are sorted by the ascending of distance from the beginning of that ray. Event points are the endpoints of the segments.

The path on this graph is the shortest possible, but because of the characteristics of the structured light cameras (the minimum allowed distance of the camera, when due to technical features the sensor can't build a cloud of points at a distance less than 50 cm), one must modify the path, adding security zones near the vertices, to avoid collisions of the robot with obstacles.

The Voronoi Diagram. The space fragmentation on the basis of a set of points on a number of areas is equal to the number of points in such a way that, in each area, any point in space will be located closer to the point from the set corresponding to that area than to any other [5].

To construct the Voronoi, partitioning of an algorithm of Fortune is used, which allows us to solve the problem in $O(n \log n)$. The method used is to sweep a line to bypass points, then at each iteration the Voronoi diagram is built only for the points before the straight line and on the straight line itself.

The path built on such a partitioning is as safe as possible on collisions, as it is always at a maximum distance from obstacles.

To perform this phase of transformation, a visibility graph was chosen. The method of constructing the graph may be less demanding on resources than discrete partitioning and, compared to Voronoi diagrams, the visibility graph represents geometric structures that simplify the creation of a route of the robot movement.

The Planning of the Robot Movement Route

The route of the robot movement can be accomplished by using several known algorithms. In particular, in robotics often algorithm A* is used [6]. Algorithm A* is a modification of the Dijkstra algorithm and based on the bypass of all points of the graph, the order of which is determined by the function $f(x)$. The function $f(x)$ is the sum of two components: $g(x)$, representing the cost to reach this node from the primary, and $h(x)$, which is the heuristic estimate of distance from a given node to the final.

The bypass goes through the following algorithm:

1. All points around the initial are recorded in a public list.
2. From the open list the point with the least value of $f(x)$ is selected.
3. If this end-point—algorithm ends, the path was found.
4. The previous point is stored in a closed list.
5. All the neighbors of the new point are inserted into the open list.
6. If the open list is empty, the algorithm ends and there is no solution.
7. Algorithm A* is used because of its simplicity and its feature of always finding a solution if it exists. In this work we selected this approach to construct the path of robot motion.

The Architecture of the Software System

For implementation of the considered image processing stages and the tasks of navigation, an architecture of the software part of the system of technical vision was developed. The solution of these tasks is executed with the help of open source libraries ROS and PCL.

A library for working with point clouds PCL (Point Cloud Library) contains objects and functions for working with point clouds in arbitrary formats. The library includes a lot of modules to solve typical tasks of computer vision. In the present work, in particular, we used modules of filtering and visualization of point clouds.

The system architecture is shown in Fig. 4. Software should perform the following functions: to receive from the structured light camera a three-dimensional point cloud; to convert the source data through the algorithms of pre-processing of the point clouds; to build a map of the environment based on a cloud with the method DBSCAN; to convert your workspace according to a visibility graph; to build a route to a given point on an obtained map with the algorithm A*.

The software contains modules that implement all of these functions (Fig. 4). The control module performs the basic management functions of the other modules and implements the following operations:

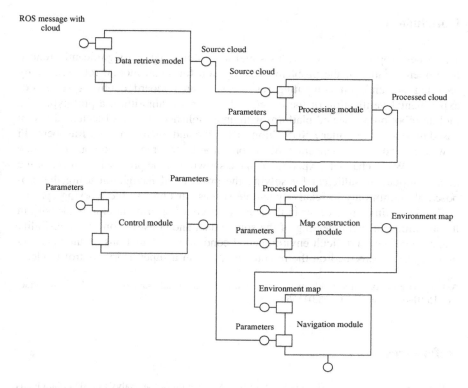

Fig. 4 Architecture of the software part of the system

- creating ROS threads for use in other modules of the program;
- receiving filters parameters from the command line or configuration file and transferring them to the processing module;
- obtaining the parameters of the map from the command line or configuration file and passing them on to the module of map processing;
- obtaining the coordinates of the current target point and transferring them to the navigation module.

To evaluate the performance of an algorithm on the constructed map of a given environment, arbitrary points on the cloud were set and, if the endpoint was reachable, we measured the time of the search algorithm to route those points obtained in the previous stages' maps. The routes to arbitrary points were successfully found on constructed maps.

On maps with lots of obstacles, the algorithm runs significantly slower due to the increased time required to build a visibility graph. According to the experimental data when the number of obstacles is more than ten, the time to decide, depending on the complexity and number of obstacles, took tens of seconds.

Conclusion

The paper proposes a set of methods and algorithms that can significantly reduce the dimensionality of the problem of navigation for small mobile robots, thereby reducing demands on computing resources of an onboard control system. For experimental verification of the developed navigation algorithm, a prototype of a mobile robot on a wheeled platform with an established system of technical vision based on sensor controller ASUS Xtion PRO Live and microcomputer Raspberry Pi 2 was created. The connection of an operator to the remote computer is done through a Wi-Fi channel. Experiments have shown that the processing power of the microcomputer is sufficient for solving the problem of navigation using the proposed algorithm only in relatively simple scenes surrounding the working space.

The large time costs occur in the stage of decreasing of point clouds' density in the algorithm of preprocessing of the source data and construction of the visibility graph, therefore, in difficult environmental conditions, calculations on the visibility graph must be executed on the remote computer in a supervisory control mode.

Acknowledgments This work was done with the financial support of RFBR (contract No. 15-01-04577-a, No. 15-08-04166-a).

References

1. Bykov SA, Leontiev VG, Skakunov VN (2012) Application of the analysis of 3D point clouds in computer vision systems. Izvestiya VolgGTU. Seriya "Aktualnyie problemy i upravleniya, vyichislitelnoy tehniki i informatiki v tehnicheskih sistemah". Vyip. 13: mezhvuz. sb. nauch. tr. Volgograd, 4(91):37–41 (in Russian)
2. Pirker K et al (2011) Fast and accurate environment modeling using three-dimensional occupancy grids. In: 2011 IEEE international conference on computer vision workshops (ICCV workshops), IEEE, pp 1134–1140
3. Jain AK (2010) Data clustering: 50 years beyond K-means. Pattern Recogn Lett 31(8):651–666
4. Borah B, Bhattacharyya DK (2004) An improved sampling-based DBSCAN for large spatial databases. In: Intelligent sensing and information processing, 2004. Proceedings of Software International Conference—IEEE, pp 92–96
5. Vasilkov MD (2011) Geometric modeling and computer graphics: computational and algorithmic foundations: a course of lectures. BSU, Minsk, p 203
6. Yao J et al (2007) Path planning for virtual human motion using improved A* star algorithm. In: Information technology: new generations (ITNG), 2010 seventh international conference. IEEE, pp 1154–1158.5

Structural and Phase Transformation in Material of Steam Turbines Blades After High-Speed Mechanical Effect

Margarita A. Skotnikova, Nikolay A. Krylov, Evgeniy K. Ivanov and Galina V. Tsvetkova

Abstract By methods of optical and electronic microscopy, tests for microhardness investigate substructure changes and the mechanisms of destruction occurring in materials of blanks. These are samples from titanium alloys PT-3V, OT-volumes by the size 100–400 μm, under which boundaries the unloading wave, connecting rotational (rotary) modes, locates adiabatic shears.

Keywords Steam turbines · Titanium alloy · Structural and phase transformation · Electron microscopy

Introduction

High-speed deformation is a modern high-efficiency way of treating metal materials. In various fields of the industry, the energy released by explosions and other methods of creating shock waves enabling manipulation of many ductile metals, including successful press forming and cutting treatments of materials. However, today in the field of mechanical engineering their remain a series of problems found in high-speed interaction of solid bodies, for example "erosion" of steam blades resulting in an effect of blows by pair drops, the intensive wear of the cutting edge of an instrument under an increase of speed and cutting treatments over a certain level. It is possible to believe that it occurs because of localization of plastic

M.A. Skotnikova (✉) · N.A. Krylov · E.K. Ivanov · G.V. Tsvetkova
Peter the Great Saint-Petersburg Polytechnic University, Saint Petersburg, Russia
e-mail: skotnikova@mail.ru

N.A. Krylov
e-mail: cry_off@mail.ru

E.K. Ivanov
e-mail: eugenei1985@mail.ru

G.V. Tsvetkova
e-mail: tsvetkova_gv@mail.ru

© Springer International Publishing Switzerland 2016
A. Evgrafov (ed.), *Advances in Mechanical Engineering,*
Lecture Notes in Mechanical Engineering, DOI 10.1007/978-3-319-29579-4_16

159

deformations in blank metal in a zone of contact with the instrument. In practice, it is most difficult to treat titanium alloys, especially, the two-phase structure of the martensite class, that is caused by their low heat conduction. We also consider high contact temperatures and a high propensity for structural and phase transformations at deformation.

Experiment

By methods of an optical metallography, transmission and scanning electronic microscopy, tests for microhardness have investited substructural changes of samples occurring in materials from titanium alloys PT-3V, OT4, VT6, VT-23, tested at speeds of deformation 10^5–10^6 s^{-1}. Tests have been executed with use of plane blanks—samples processed by a shock wave with the help of a pneumatic gun, explosion and a cutting tool.

Cutting Treatment of Titanium Alloys

The treatment of titanium alloys PT-3V and VT-23 was carried out by a hard-alloy cutter T15K6 without lubrication with a feed speed $S = 0.26$ mm/revolutions and cutting depth $t = 3$ mm, in a range of cutting speeds 2–120 m/min. The geometrical parameters of a cutter made corners: $\varphi = 45°$; $\varphi_1 = 15°$; $\alpha = 6°$; $\gamma = 12°$.

Morphology of Chip

As it is visible from Fig. 1a, at increase of speed of a cutting treatment blank from titanium alloy VT-23, the linear wear of the tool was much higher, in comparison with treatment of steel HVG or aluminum alloy AMz. Thus chip local (adiabatic) shear was formed with traces of localization of plastic deformation (ε_{local}), Fig. 1b. The period of localization (width of chip segments) is on the average 300–400 μm.

Structure of Chip Metal

The structure of blank from titanium alloy VT-23 in an initial state representing colonies of parallel plates of α-phase and disjointed interlayers of β-phase, Fig. 2a. At cutting treatment, beginning already from speed 2 m/min, the inhomogeneous plastic deformation, its strong localization in narrow iteration with a period of 300–400 μm volumes of metal on the mechanism of formation of a superfine structure, Fig. 2b, was observed.

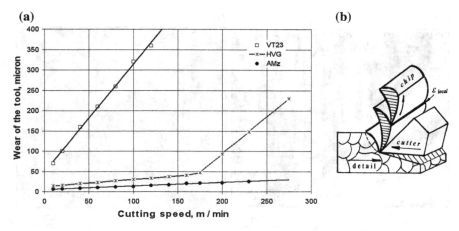

Fig. 1 Wear of tool with increase of speed cutting treatment of alloys VT-23 (*1*), HVG (*2*), AMz (*3*), (**a**). The scheme of «adiabatic shift» chip formation (strong localization of plastic deformation) (**b**)

We have shown some results of using a scanning electronic microscopy. Free surfaces of chips from alloys PT-3V (Fig. 2c) and VT-23 (Fig. 2d) were formed on a rotational mechanism with attributes of destruction in local adiabatic shear conditions.

Estimate of Chip Microhardness

Microhardness testings were made in chip metal from alloy VT-23 along a direction of movement of a cutter with speed 120 m/min and an interval of 20 μm at loading of 20 g. Results of microhardness testing of chips had wavy character and is especially close to their free edge, Fig. 3. The maximal values of microhardness had on places of an articulation of segments of chips, in which with the help of transmission electronic microscopy the localization of plastic deformation on the mechanism of formation of narrow zones has been earlier found out by a superfine structure. Here absolute values of microhardness for alloys VT-23 reached 4381 MPa at average hardness of chips and metal in an initial condition, especially to 3761 and 3903 MPa.

It is necessary to note that the specified changes of structure, Fig. 2b and modulation of microhardness, Fig. 3, near to the free edge of a chip, were more essential than in a cutter zone that allowed it to have more intensive relaxation processes.

Thus, the cutter at the movement along treated blank material forms a wave of compression which modulatesa material structure, dissecting it on mezo-volumes by the size 300–400 μm. Having reflected from a surface of blank material, the unloading wave of the plastic deformation that provides a connection of rotational (rotary) modes of plastic deformation and locating adiabatic shears along the

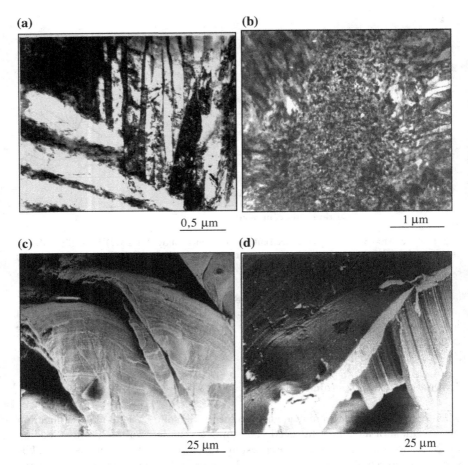

Fig. 2 Structure of chip from alloy VT-23 before (**a** and **c**) and after (**b** and **d**) cutting treatment with speed 120 m/min

boundaries of educated mezo-volumes is formed. Dissipative modulation of structure and microhardness in titanium alloys could be the reason of decrease of tool wetarproofness.

Shock Stressing of Titanium Alloys

The treatment by a shock wave was carried out in a material of plane blanks—samples from two-phase ($\alpha + \beta$) titanium alloys OT4, VT6 and VT-23, tested by anvil block, or blast wave [1–4]. Thus blanks have been finished with full destruction of two free surfaces or cavities.

Morphology of Destruction of Blanks Material After Shock Stressing

On Fig. 4 we submit a photo and the scheme of destruction of a blank with a sample of alloy VT6, by a treated shock wave with speed 568 m/s. It can be seen that in the formed turnpike the crack was parallel to a free surface of a target and had the step form. It is possible to allocate three stages (3 zones) of destructions.

Zone 1—"epicenter" of shock wave in diameter no more than 15 mm, in which characteristic viscous "tunnels" focused along a direction of impact were formed. A zone 2—ring-shaped "periphery" of a shock wave. Probably, in this zone was transversal spreading of a wave of compression to loss of speed and energy. A zone 3—"final destruction".

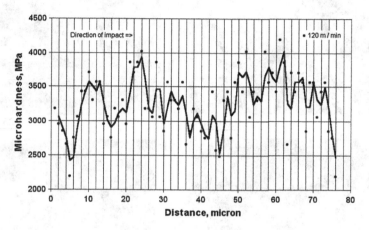

Fig. 3 Distribution of microhardness in chip metal from alloy VT-23 along a direction of movement of a cutter with speed 120 m/min

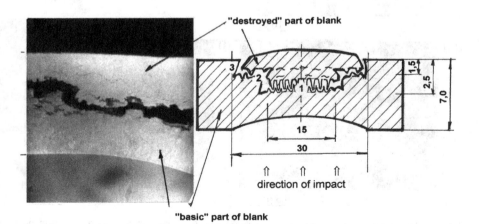

Fig. 4 Photo of share section "destroyed" and "basic" of parts of the blank. The scheme of destruction of plane metal blank at shock stressing with a speed 568 m/s. Sizes are specified in mm

There were cases in surfaces of breaks having at a macro-level antisymmetric character, at micro-level that did not coincide. This required disposal of material, which is especial in a zone 2, Fig. 4. After termination of testing, the surface "destroyed" parts appeared straightened, (the internal stresses were removed), and the "basic" part of the blank retained a strong camber.

On Fig. 5a–f electron microscope photos of breaks and cross-section "destroyed" and "basic" of parts of the blank in zones 1 and 2 are submitted. It can be seen that

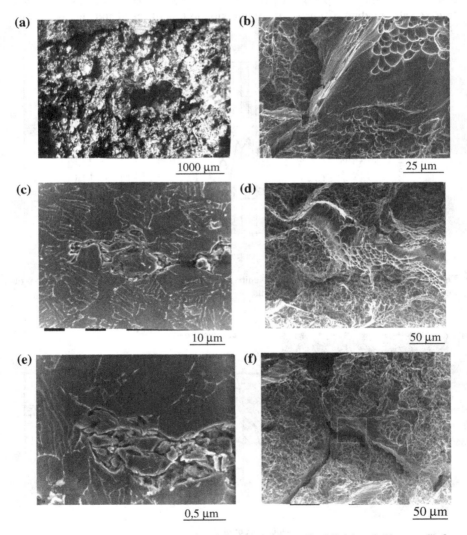

Fig. 5 Microstructures in cross-section and fracture surfaces "basic" (**a**) and "destroyed" (**b–f**) parts of a target in a zone 1 (**b–d**) and in a zone 2 (**e–f**), up to (**a, d, f**) and after (**b, c, e**) the etchings

in a break of the destroyed blank on mezo-level were formed components of a breaking-off round form by the size 100–200 µm (Fig. 5a, d) and 300–400 µm (Fig. 5f), accordingly, in a zone 1 and in a zone 2. Such mezo-volumes have been internally broken on more shallow micro component, too round forms 10–40 µm (Fig. 5c, e). Such viscous sites of metal of the destroyed blank could be generated as a result of the rotational mechanism of plastic deformation along narrow micro channels, which direction coincided with a direction of operation of the maximal stress. Along the boundaries of such channels the superfine, structural single-phase condition was formed (in an initial condition of a two-phase alloy), as etching of fracture surface in such sites, presence of the second β-phase component has not revealed, (Fig. 5b).

It is possible to believe of blank material that, breaking it on mezo-volumes by the size 100–400 µm, (Fig. 5d, f) under which boundaries the unloading wave, connecting rotational (rotary) modes, makes located adiabatic shears (Fig. 5b) and destruction.

Estimation of Microhardness of Blank Metal After Shock Stressing

On Fig. 6 results of microhardness testing with an interval of 20 µm are submitted at loading of 20 g, in blank metal from alloy VT6 after treatment by a shock wave with speed 568 m/s. Measurements have been executed starting from an edge of a "basic" blank part on a trace of movement of wave in the central and peripheral field on a distance of 4 mm from the center, in both cases at level of zone 1. Absolute values of microhardness in the center and on periphery were reached on the average 4413 and 3996 MPa, accordingly, at average hardness of sample metal in an initial condition 2416 MPa.

Apparently from Fig. 6, results of measurement of microhardness after shock stressing had a wavy character with the size of a half wave 100–200 µm. In comparison with central, the peripheral wave on distance of 4 mm from the center was in an antiphase and thus, they were self-consistent in mezo-volume 100–200 µm (zone 1).

Similar comparative results have been received and at level of zone 2. Peripheral wave on distance of 11 mm from the center, was self-consistent with central less often, but the size of such volumes increased up to 300–600 µm.

Structure of Blank Metal After Shock Stressing

Structure of blanks metal from alloys OT4, VT6 and VT-23, tested anvil block and a blast wave investigated with the help of transmission electronic microscopy, Fig. 7. Results have shown that, with an increase of speed of a shock wave, localization of plastic deformation took place, to which decomposition of enriched solid solutions always preceded. In all investigated materials in boundary layers generated mezo-volumes, interlayers of β-solid solution under operation of shock wave, were exposed to fractional or decomposition (dissolution) resulting in β → α-transformation. In such places the structure from fine grains of different orientation which had a heightened microhardness was formed and yielded in condition of microdiffraction "ring" electronogrames, Fig. 7a, an origin of micro-cracks here was observed.

As it has been shown in earlier works [5–8], with increase of coefficients of β-stabilization of titanium alloys, decomposition of layers of β-solid solutions, with a composition appropriate to them, there develops in temperature-time intervals corresponding to them, that at lower temperatures and for longer time, then they are more enriched with the same β-stabilizing alloying elements. Therefore with increase of speed and reduction time of shock stressing, the β-interlayer of more alloyed alloys VT-23 and VT6 appeared to be steadier against decomposition in

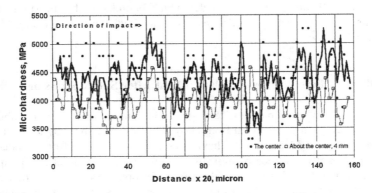

Fig. 6 Distribution of microhardness in blank metal from alloy VT6 along a direction of movement of shock wave in a zone 1 with speed 568 m/s

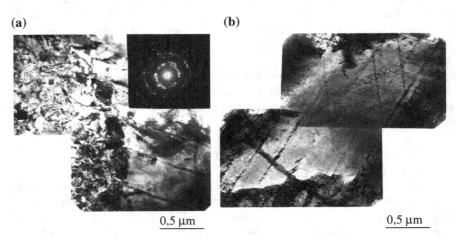

Fig. 7 Electron microscope structures after treatment by blast wave of blank metal from alloy OT4 (**a**) and VT-23 (**b**)

comparison with alloys PT3V and OT4. As can be seen from Fig. 7, at identical stressing by blast wave in blank from alloy OT4 (Fig. 7a) the localization of plastic deformation was observed, while in alloy VT-23 the decomposition of β-interlayer (Fig. 7b) took place only, previous to it.

Conclusion

It is possible to believe that, at high-speed machining, the compressing shock wave modulates material structure, breaking it on mezo-volumes by the size 100–400 μm. Inside formed mezo-volumes, the phases of waves are opposite in sign, that results in a relative relaxation in them of stresses.

As a result of self-organizing of a system, the unloading wave of plastic deformation and destruction, depending on relaxation ability of material (structural and concentration [3, 5, 8], energies of defect of packing, ability of transformation of mechanical energy in thermal, realization of phase transformations) is formed.

At micro-level the wave of plastic deformation forming, making multiple copies and self-organizing defects of crystal structure, calls a strain hardening micro-volumes of blank metal. For saving stability of deformable material, the speed of strain hardening should predominate the above speed of dynamic structural and concentration relaxation. At falloff of material ability to strain hardening on a micro-level, in plastic deformation metal volumes on mezo-level are involved.

At increase of speed machining of blanks from titanium alloys, having a low heat conduction, the speed of strain hardening at the expense of transmitting modes of plastic deformation on micro-level quickly decreases (is braked) and it appears commensurable with speed of relaxation weakening. A surplus of thermal energy results in its redistribution (localization) and to development of plastic deformation in which volumes of metal on mezo-level are involved. Thus the rotational modes of plastic deformation, its resulting in localization and destruction are connected. In titanium alloys of martensite class in which, as is known, plastic deformation accelerates decomposition enriched β-solid solution with formation α''-phases [4]. Its subsequent ageing at temperatures 450–500 °C results in phase $\alpha'' \rightarrow \alpha' + (\beta)$—transformation and in significant hardening alloys VT6 and VT-23 [9, 10]. Connection of new modes of plastic deformation on the mechanism of phase transformations, considerably complicates achievement of exhaustion of the plasticity necessary for destruction of material at shock stressing.

References

1. Mescheryakov Y, Divakov A (1989) In: The Pre-print, RAS, St. Petersburg, Russia, p 36
2. Skotnikova MA, Chizhik TA, Lisyansky AS, Simin ON, Tsybulina IN, Lanina AA (2009) Research of working shovels of turbines of big power taking into account structural and phase transformations in metal of stampings from a titanic alloy of VT6. Metalloobrabotka 54(6):12–21

3. Skotnikova M, Ushkov S (1999) In: Proceedings conference titanium—99, St. Petersburg, Russia, vol 1, p 414
4. Skotnikova MA, Vinogradov VV, Krylov NA (2005) The accounting of the wave theory of plastic deformation at high-speed machining of surfaces of preparations. Metal working, SPb, №11 pp 12–15
5. Skotnikova M (1997) In: Proceedings conference MORINTECH-97, St. Petersburg, Russia, vol 4, p 251
6. Grindnev V, Ivasishin O, Oshkaderov S (1986) In: The book, Kiev, Ukraine, p 256
7. Yrmolov M, Solonina O (1967) PMM, Moscow, Russia, vol 23(1):63
8. Skotnikova MA, Zubarev YM, Chizhik TA, Tsybulina IN (2004) Structural-phase transformation in metal of blades of steam turbines from alloy VT6 after technological treatment. Proceeding of the "10th world conference on titanium" 13–18 Jules 2003, Hamburg, Germany, vol 5, pp 2991–2999
9. Skotnikova MA, Krylov NA, Mescheryakov JI, Radkevich MM, Ivanov EK, Mironova EV (2011) Formation of rotation in titanium alloys at shock loading. In: 12th World conference on titanium—2011. Proceedings of the conference of the nonferrous metals society of China, held in Beijing, China, June 19–24, pp 540-543
10. Skotnikova MA, Krylov NA, Tsvetkova GV, Ivanova GV (2013) Structural and phase transformation in material of blades of steam turbines from titanium alloy after technological treatment. Advances in mechanical engineering. Selected contributions from the conference "modern engineering: science and education", Saint Petersburg, Russia, June 20–21, pp 93–101

Stress Corrosion Cracking and Electrochemical Potential of Titanium Alloys

Vladimir A. Zhukov

Abstract This paper considers the relationship of electrochemical parameters and corrosion resistance stress cracking (SCC—Stress corrosion cracking) of titanium alloys in artificial seawater. It is established that there is a correlation between the decrease of the resistance SCC low-cycle fracture and the change of the potential of oxygen evolution.

Keywords Titanium alloy · Stress corrosion cracking · Potential of oxygen evolution · Low-cycle fatigue strength · Thermo-EMF temperature coefficient

Introduction

Despite the relatively high cost of materials and manufacturing technology designs, the field of application of titanium alloys in various branches of engineering continues to expand [1]. Research and development actively are conducting with the aim of using these alloys as material of pipelines in offshore oil and gas [2], and structural elements of the loaded parts of offshore platforms.

One of the most significant barriers to the use of medium- and high-strength alloys based on titanium is a sharp decrease of resistance of stress corrosion fracture in aqueous solutions of chlorides in the presence of an acute stress concentrator or cracks, first identified in large-scale tests to determine the critical stress intensity factor [3]. SCC is found in various industries: aerospace machinery, nuclear reactors, engineering services and pipelines [4]. The influence of various factors of stress corrosion cracking under tension is usually estimated by the change of K_{ISCC} on testing flat specimens with a fatigue crack. The specimen thickness W must conform to the requirement $W \geq 2.5K_I^2/\sigma_Y^2$ [5]. The specimens with fatigue crack were tested

V.A. Zhukov (✉)
Peter the Great Saint-Petersburg Polytechnic University,
St.-Petersburg, Russia
e-mail: v.zhukoff2011@yandex.ru

also with active stretching by low-speed movement of the grips of the testing machine, for example from 0.02 to 0.005 mm/min [6]. The tests under low-cycle loading are sparse. The greatest effect SCC of titanium alloys is observed at loading frequencies of the order of cycles per minute and that effect decreases at lower frequencies or increases at the duration of the test [7].

Despite considerable interest in the problem of SCC, the concern of estimating the sensitivity of the material of the workpieces and welding elements of structure manufactured of titanium alloys by the action of corrosive environment is still relevant. This is because even the laboratory test by X-ray structure analysis does not detect areas of structural precipitation in titanium alloys [8] which tend after aging for SCC in aqueous solutions of chlorides. The most sensitive methods of evaluation of structural transformations in the process of aging alloys are methods of electrical resistivity and methods of temperature coefficient ε_t of thermo-EMF. However, these methods are practically useless if there is a neccessity to appreciate the structural state upon completion of the technological process or during working of the equipment.

It is natural to include the parameters of the electrochemical interaction of a metal with the solution in a group of values thereby which the investigator may be judge the resistance of titanium alloys by "rapid" stress corrosion fracture. This paper presents the results of conformity of electrochemical potential of titanium alloys in aqueous solution and the effect of SCC.

Experimental Results

Figure 1a presents the dependences of the relation between long-time strength by static bending of β-alloy Ti-10 V-11Cr-3Al (B-120VCA) flat specimens with a fatigue crack or a sharp notch and the breaking strength F_f of similar specimens by short-term static bending in the air. Figure 1b shows the dependences of the relation between the low-cycle fatigue strength of cylindrical specimens with sharp notch by the asymmetry factor $R \approx 0$ and the breaking strength F_f of similar specimens by the short-term static tension on the air. The effect of the negative influence of 3.5 % aqueous NaCl solution by the low-cycle fatigue of cylindrical specimens with the sharp notch were found more acute than by the long-time strength of the bending specimens with a crack. Based on these and similar data of other titanium alloys, the authors of the article [9] proposed to conduct tests by the low-cycle fatigue in a corrosive environment of the small size cylindrical specimens with the sharp notch. It is possible to significantly reduce the cost of testing compared to testing the specimens with a crack.

Resistance to low-cycle fracture of the specimens with a sharp notch is evaluated through the estimated plastic deformation value of the submicroscopic volume, located with the crack tip at a distance equal to the low-cycle fracture pitch in the past (before the failure of this volume) cycle of deformation $\varepsilon_{f\sigma}$ [10]:

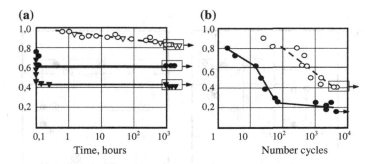

Fig. 1 The effect of 3.5 % aqueous NaCl solution on the long-time strength by static bending (**a**) and on the low-cycle fatigue strength (**b**) alloy Ti-10 V-11Cr-3Al (B-120VCA): ○ ▽—tests on the air; ● ▼—tests in the solution; ○ ●—specimens with the sharp notch; ▽ ▼—specimens with the fatigue crack

$$\varepsilon_{f\,\sigma} = \frac{K}{E\sqrt{6\pi v}},$$

where K is the magnitude of the stress-intensity factor at the asymmetry factor $R \approx 0$ by the low-cycle fatigue; E—the Young's modulus of the material; v—the step of crack at that K.

In accordance with the conditions similarity, test materials with various values of tension yield σ_y^t were realized under the condition $K/\sigma_y^t = idem$, or $r_y^{1/2} = idem$, where r_y is the radius of the plastic zone at the crack tip in Irwin G.R. at a given value of the stress-intensity factor. The number of samples in a series was in a range of 4–10. The potentials of oxygen evolution V_O and hydrogen evolution V_H were determined by polarograph at automatic voltage changes from −0.9 to +0.9 V; the surface of the specimens were damaged by a glass scraper in the solution under voltage −0.9 V.

The relationship between the electrochemical properties of titanium alloys and critical local deformation $\varepsilon_{f\,\sigma}$ in solution at $r_y^{1/2} = 0.055$ mm is presented in Fig. 2. All alloys were exposed to plastic working within β-region and aging at 773 K. The dotted line in Fig. 2 corresponds to the lower boundary $\varepsilon_{f\,\sigma} = 0.09$ of confidence interval at p = 0.05 for alloy ПТ3B tested in air.

In the state after hot working pressure with air cooling commercially pure titanium and alloy ПТ3B from 4.5 % Al, and as well as α + β-alloys containing 6 % Al after the quenching from β, the region is practically not sensitive to the action of a corrosive environment. Following after the quenching, the aging of alloys with high aluminum content leads to a significant decrease of the resistance of corrosion-mechanical failure.

It can be noted that for the same materials such as alloys BT6 (8−11) and BT20 (12−14) a match occurs between the value $\varepsilon_{f\,\sigma}$ and value potential of oxygen evolution. For α- and α + β-alloys, the value $\varepsilon_{f\,\sigma}$ which is in the area of greater than

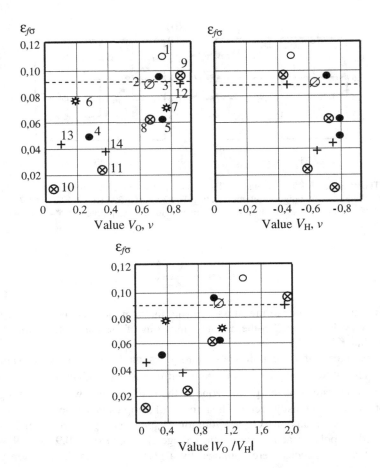

Fig. 2 The dependence $\varepsilon_{f\sigma}$ of the titanium alloys in aqueous solution NaCl: ○—BT-1; Ø—ПТ3В; ●—BT6; ✿—В-120 VCA; ⊗—BT6; +—BT20; 1, 2, 5, 6, 7, 8—air cooling; 4—cooling with the furnace; 3, 9, 12—quenching, 10—air cooling + ageing; 11, 13, 14—quenching + aging

0.09, the potential of oxygen evolution is not less than 0.65, and the ratio of potentials of oxygen and hydrogen is not less than 1.1. The potential of oxygen evolution less than 0.4 V shows about significant decrease of the corrosion-mechanical failure resistance of α + β-alloys.

This correspondence between the potential of oxygen evolution and prone to the corrosion-mechanical failure allows conclusion about the significant influence of structural state on the capability of titanium alloys to the restoration of the protective barrier layer destroyed during deformation.

Figure 3 shows the change of electrochemical potential of alloy BT-1 and alloy BT6 in 3.5 % aqueous NaCl solution. The measurements were carried out on the

Fig. 3 Change EHP after
failure of specimen: BT1 and
BT6 (aging)

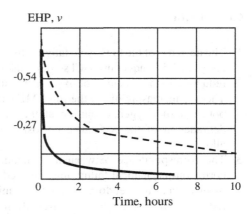

specimen which after its destruction by a tensile testing machine remained in the
same container with the solution. The recovery rate of the barrier layer of the alloy
with a lower value of the potential of oxygen evolution is significantly less.

To evaluate the correspondence between the parameters of the electrochemical
interaction in the system "metal–solution" and the tendency of titanium alloys to
corrosion-mechanical failure it was used thermo-EMF (thermocouple voltage)
temperature coefficient method for the comparison of different structural states
during of the ageing. The results of the study are presented in Fig. 4. The dependence of the thermo-EMF temperature coefficient ε_t of titanium alloy BT20 is
similar the dependence of the specific electrical resistance of aluminum and other
alloys by the duration of aging. The first section of this dependence corresponds to
the formation of local inhomogeneities on the structural imperfections of the crystal
lattice. It is at this stage of aging reveals a sharp decrease in the fracture resistance
of titanium alloys and the potential of oxygen evolution. The reduction potential of
oxygen evolution means that the capability of adsorbing oxygen from the solution
thus forming the barrier layer in the process of deformation is reduced.

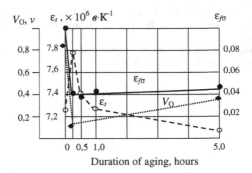

Fig. 4 The dependence of the thermo-EMF temperature coefficient ε_t and the potential of oxygen
evolution V_O with duration of aging of alloy BT20

Conclusion

1. The decrease of the low-cycle fatigue strength of the pseudo-α- and α + β-titanium alloys in 3.5 % aqueous NaCl solution takes place with a simultaneous significant reduction in the potential of oxygen evolution.
2. There is increasing of the thermo-EMF temperature coefficient and reducing the potential of oxygen evolution at the primary stage of aging, causing quick degradation of the resistance corrosion-mechanical failure of α + β-titanium alloy.
3. The correspondence between the potential of oxygen evolution and prone to corrosion-mechanical failure may be used for detection directly on the structure of such states as pseudo-α- and α + β-titanium alloys which are characterized by reduced corrosion-mechanical strength in aqueous solutions of chlorides.

References

1. Filippov GA, Licyansky AC, Nazarov OI, Tomkov GP (2008) The tendency of perfection of high-speed steam engines for nuclear power stations. Energ Mach Equip (Powermachinebuilding, Saint Petersburg) 3:3–12
2. Gorynin IV, Ushkov SS, Hatunchev AI, Loshakova NI (2007) Titanium alloys for marine engineering. Poletechnika, Saint Petersburg, 387 p
3. Brown BF (1966) Material research and standards 6:3–129
4. Raja VS (2011) Stress corrosion cracking. Theory and practice. Bombay, India, 816 p
5. Dietzel W (2001) Fracture mechanics approach to stress corrosion cracking. Anales de mecanica de la fractura (18)
6. Barella S, Mapelli C, Venturini R, Investigation about the stress corrosion cracking of Ti-6Al-4 V. [Electronic resource] http://www.arcam.com/wp-content/uploads/Arcam-Ti6Al4V-ELI-Titanium-Alloy.pdf
7. Dawson DB, Pelloux RM (1974) Corrosion fatigue crack growth of titanium alloys and aqueous environments. Met Trans 5(3):723–731
8. Zwicker W (1979) Titanium and its alloys. In: Elutin OP, Glazunov SG (eds) Lang Trans with German. Metallurgy, Moscow, 512 p
9. Zhukov VA, Ivanova LA, Marinech TC, Razuvaeva IN, Hesin YD (1981) Thermal stability of the pseudo-α-titanium alloys and methods of its estimation. Metal Sci Heat Treat Metals (Moscow) 12:37–39
10. Zhukov VA (1994) Quantum mechanical approach to the analysis of effects in deformation and fracture of metals. Metals (Moscow) 6:93–97

Metal Flow Control at Processes of Cold Axial Rotary Forging

Leonid B. Aksenov and Sergey N. Kunkin

Abstract Axial rotary forging, as local methods of metal forming, allows expanding opportunities of technological processes at smaller power equipment. This paper studies the processes of axial rotary forging for manufacturing of different parts from tube-blanks on machines that are used for forming a conical roll with angle of inclination 10–25° or two cylindrical rolls, i.e. rolls with tilt 90° to the axis of the blank. Possibilities of rotary forging are limited due loss of stability of tube-blanks which happened during forming of wide and thick flanges. However, there are not too many ways to control metal flow in this process. It is considered the main factor in determining the direction of flow of metal that produces friction force on the contact surface of formed metal with forging rolls. Direction of this force can be changed by the position of forming rolls with respect to the blank. The technology of axial rotary forging with displaced forming rolls provides a stable forming process of wide collars and thick flanges.

Keywords Axial rotary forging · Metal flow · Forging roll · Displacement of rolls · Contact area · Friction force · Outward flanging · Flanges · Collars

Introduction

Large quantities of parts such as flanges and collars are used in a variety of industries. Production of flange parts is performed with different technologies, but they do not have a high utilization rate of metal. Many production technologies of flanges are based on the technology of hot forging with subsequent further processing [1]. In this case, the plasticity of the metal is higher and forming occurs with small technological force. However, the hot deformation processes require

L.B. Aksenov (✉) · S.N. Kunkin
Peter the Great St. Petersburg Polytechnic University, St. Petersburg, Russia
e-mail: l_axenov@mail.spbstu.ru

S.N. Kunkin
e-mail: kunkin@spbstu.ru

© Springer International Publishing Switzerland 2016
A. Evgrafov (ed.), *Advances in Mechanical Engineering*,
Lecture Notes in Mechanical Engineering, DOI 10.1007/978-3-319-29579-4_18

significant expenditure, particularly on the parts covered with scale which requires additional machining. Therefore, the application of these processes is not very effective in industry. A cold axial rotary forging has wider advantages as it does not require heating and is characterized with high accuracy and good quality of surface parts. Naturally, in cold forming, the required technological force will be stronger than during hot forming, and it lowers the plasticity of metal, which makes higher demands on their analysis [2, 3].

Axial rotary forging technology is intended for manufacture of different axisymmetric parts from bars or hollow blanks [4–6]. This technology is representative of other processes with local deformation of worked metal. While in contact with forming tools is only part of the metal blank, which reduces the square of the contact area, maximum of the contact stresses, and, accordingly, the required forming force. This provides certain advantages to this process in comparison with other metal forming technologies, for example, hot forging [7].

Machines Used for Cold Axial Rotary Forging

The processes of axial rotary forging on two machines has been studied. One of them used a conical forming roll with an angle of inclination of 10–15° (Fig. 1a). At the other machine the process is used to form two cylindrical rolls, i.e. rolls with a slope of 90° to the axis of the workpiece (Fig. 1b).

The simplest type of rotary forging machine is a machine with a drive for blank rotation and passive rolls receiving the rotation from the blank due to friction forces on the contact surface. Machines that work by such action have a number of advantages:

- the centre of mass of the tool is not rotated about the vertical axis and, therefore, the deformation process can be greatly intensified by increasing the speed of rotation of the blank and the blank deformation per revolution;
- reduced demands to the stiffness of the bed and to the mass of foundation;
- reduced noise levels and also risk of damage for machine elements due to the reduction of low and high frequency vibrations.

Axial motions of forming rolls and blanks along the axis of blank rotation determine the feed and degree of metal deformation per one revolution of tube-blank. It is carried out under the action of axial forces in different ways: motion only of forming rollers, the motion only of formed blank or synchronous motion of rolls and blank towards each other.

A rotary forging machine consists of two main technological units—upper and lower. The lower unit contains the spindle with a drive for blank rotation, bearing units and pusher. The upper unit has a drive for transverse displacement of the roller relative to the blank axis and a drive for rotation of rolls.

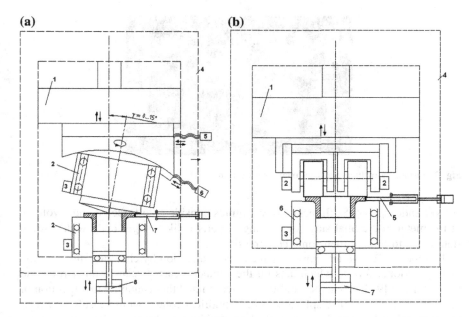

Fig. 1 Schemas of machines for cold axial rotary forging: **a** with conical roll (*1* ram, *2* upper unit, *3* drive for blank rotation, *4* press frame; *5* drive for forming roll, *6* drive to change angle of inclination of conical roll, *7* cross roller, *8* pusher); **b** with two cylindrical rolls (*1* ram, *2* upper unit, *3* drive for blank rotation, *4* press frame; *5* cross roller, *6* upper unit, *7* pusher)

A drive for forming rolls is necessary when the upper unit of the rolls (including the spindle and bearing units) has a large mass and inertia. In the case of rotary forging flanges made of thin-walled tubes (for example, with wall thickness S = 3 mm and less) at the initial moment of outward contact area between forming rolls and the blank has a minimum value and friction forces on the contact area are small. This phenomenon can lead to negative consequences: the process of rotary forming becomes impossible as the blank loses its stability and is crumpled by the forging roll due to lack of friction forces for implementation of outward flanging at the early stages of process (Fig. 2).

Control of Metal Flow

Possibilities to keep metal flow at rotary forging under control are very limited. Different types of applicable tools (cross-rollers, forming rollers with shoulders, mandrels) can limit the metal flow in certain directions, and after the forming of some area of the part to redirect the metal in the place, where the forming of a part is

Fig. 2 Instability of tube-blank at axial rotary forging

not yet ended [5–8]. Such technology requires considerable force, as most volume of the metal in the final stage of forming of parts is like a rigid body with a strain state close to three-dimensional compression.

It is more effective to change the metal flow direction by changing the direction of the frictional forces, which acts on the contact area between forming rolls and formed metal [9, 10]. The possibility was determined that changing the direction of the friction forces with displacement of cylindrical rollers would result in a relatively traditional layout. So to receive the outer flanges, the metal flow on the outer face of the blank is required. For the direction of friction forces in the required direction at rotary outward, flanging cylindrical rollers should be placed with some displacement δ relative to the transverse axis of the tube-blank (Fig. 3).

Magnitude and direction of friction forces on the contact area at the rotary forging of axisymmetric parts with cylindrical rolls are determined by the sliding speed of the roll's points relative to the surface of the blank. Along with slipping of deformed metal, relative forming rolls cause kinematic characteristics of an outward rotary process, with metal slips also on the contact area in the tangential and radial directions due to motion of formed metal. As a result at the contact area, components of the velocity are imposed on each other and, depending on its sign, will be added or subtracted.

The offset of the rolls leads to an increase in velocity component of movement of the metal is directed from the centre of the workpiece. This phenomenon is difficult to explore on the basis of the simulation process, as the film speeds in this case shows the complete velocity vector, in which the main component is caused by the rotation of the workpiece [10].

Similarly, for operation of the outward-flanging from tube-blank with conical roll, it is recommended to set the top of the conical roll, not on the axis of symmetry of the workpiece, and with an offset of $\delta = (0.1 - 0.3)R$ (Fig. 4).

It should be noted that the tangential kinematical slip is a few times greater than the slip caused by a deformation of blank in the tangential direction. Thus, there are

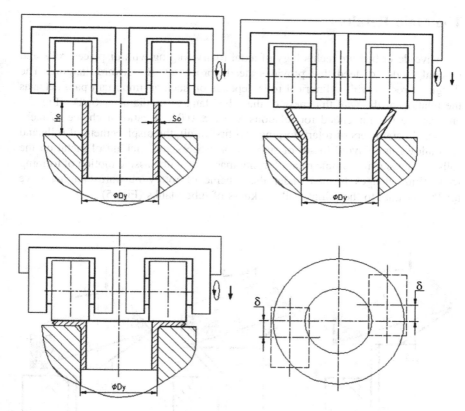

Fig. 3 The rotary outward-flanging by cylindrical rolls with displaced axes

two zones with different conditions of friction on the contact area at cold axial rotary outward flanging with cylindrical rollers:

1. Ahead zone, located on the flange on the outside of the diameter of the engagement. Here the slide of metal in the direction of rotation of a forming roll due to the fact that a blank moves faster than the rollers on this part of the contact area.
2. Lag zone, located on the inner side of the diameter of engagement. In this zone a metal slide along a contact area is determined by the lag of the blank from the forming rolls.

The ability to control the forces of friction on the contact surface expends the technological possibilities of a rotary forging, helps to reduce the value of contact stresses and has positive influences on the process of metal flow and the forming of flanges.

The Main Results

We have described the technology of axial rotary forging with displaced rolls and rational modes of feed that provides steady processes of outward-flanging. The flanging process with cylindrical rolls depends on such an important parameter as the relation of the wall thickness of the tube-blank to its diameter S_0/D_y (Fig. 3), and can be recommended for relations $S_0/D_y \leq 0.04$. A rational choice of technological parameters of rotary forging and first of all, the displacement of rolls and feed allows us to avoid losses of stability of process and sticking of metal on the rolls. It allows using simple rotary forging machines with passive motion of forming rolls. Wide flanges formed under the scheme of rotary outward-flanging have thickness equal to the initial wall thickness of tube-blanks (Fig. 5).

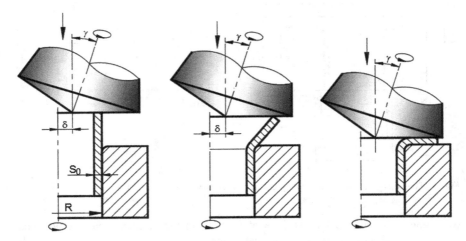

Fig. 4 The rotary outward-flanging by conical roll with displaced axe

Fig. 5 Rotary forged flanges from tube-blanks with displaced conical roll

Resume

- An effective way to control the flow of metal at processes of axial rotary forging is to change the direction of the friction forces at the contact surface, achieved by a particular arrangement of the forging rolls: non-axial position of the cylindrical rolls and offset conical roll relative to the axis of the workpiece.
- The technology of axial rotary forging with displaced forming rolls provides the stable forming processes of:

  - outward-flanging of collars with width over diameter of the workpiece and final thickness equals to initial wall thickness of tube-blanks;
  - flanges with thickness more than 2.5 wall-thickness of tube-blank.

Acknowledgements This research was sponsored by company "JA-RO AB" (Finland).

References

1. Semenov LP, Semenov AA, Pasko AN (2010) Shaping of the outer walls thickening at tubular billets (in Russian). Kuznechno-shtampovochnoe proizvodstvo. Obrabotka metallov davleniem #9:33–37
2. Gurinovich VA, Balandin YA, Gurchenko PS, Kolpakov AS, Zharkov EV, Isaevich LA, Sidorenko MI (2005) Axial rotary forging of flange type parts (in Russian). Avtomobilnaja promjschlennost #9
3. Surkov CA, Koryakin NA, Galimov RE (2005) Stamping by orbital forging of flange and ring blanks (in Russian). Zagotovitelnoe proizvodstvo v mashinostroenie #7:27–29
4. Trituhin VV (2010) Closed die forging method of upsetting with extrusion (in Russian). Kuznechno-shtampovochnoe proizvodstvo. Obrabotka metallov davleniem #6:43–45
5. Aksenov LB, Kunkin SN, Elkin NM (2011) Axial rotary forging of flanges for tube connectors (in Russian). Metalloobrabotka Research-and-production journal #3(63):31–36
6. Aksenov LB, Kunkin SN (2013) Manufacturing technology of rotary forging of ax symmetric parts with flanges. Modern engineering. Science and education. In: Materials of the international scientifically-practical conference. Publishing house-Polytechnic University, Saint-Petersburg, pp 858–866
7. Nowak J, Madej L, Ziolkiewicz S, Plewinski A, Grosman F, Pietrzyk M (2008) Recent development in orbital forging technology. Int J Mat Form 1:387–390
8. Groche P, Fritsche D, Tekkaya EA, Allwood JM, Hirt G, Neugebauer R (2007) Incremental bulk metal forming. Ann CIRP 56:635–656
9. Levanov AN (2013) Method of tests of contact friction in processes of metal forming (in Russian). Kuznechno-shtampovochnoe proizvodstvo. Obrabotka metallov davleniem #3:43–47
10. Han Xinghui, Hua Lin (2012) Friction behaviors in cold rotary forging of 20CrMnTi alloy. Tribol Int 55:29–39

Use of the Capabilities of Acoustic-Emission Technique for Diagnostics of Separate Heat Exchanger Elements

Evgeny J. Nefedyev, Victor P. Gomera
and Anatoly D. Smirnov

Abstract Interpretation of AE testing results of large-size vessels and units is usually a non-trivial task. The higher its complexity, the higher is a more complex design of the object under testing. In this respect, shell-and-tube heat exchangers are one of the most complex objects. The paper deals with methodical techniques for evaluation of AE testing data, which provide increasing the confidence of testing by combining various location algorithms and evaluation of the results of their use. The suggested approach implements the possibility of defect detection in various structural elements—both housing material and tube bundles.

Keywords Acoustic emission · Heat exchanger · Planar and 3D location

Introduction

Acoustic emission is a modern method of nondestructive testing [1–3]. The method is quickly spreading due to its properties such as high sensitivity, the ability to record developing (the most hazardous) defects, integrity, and, consequently, high performance. Nowadays the AE method is well-developed for crack detection in cases of static failure, cyclic failure and leakage detection [4–7]. The basic techniques for diagnostics of pressure vessel housing parts were elaborated [8–10].

The complexity of interpretation of AE testing results of large-size vessels and units is in many ways related to the need for separation of the most useful component from the large volume of recorded signals—the data related directly to the

E.J. Nefedyev (✉)
Central Boiler and Turbine Institute (CKTI), Saint-Petersburg, Russia
e-mail: ne246@ya.ru

V.P. Gomera · A.D. Smirnov
Kirishinefteorgsintez, 187110 Kirishi, Russia
e-mail: Gomera_V_P@kinef.ru

A.D. Smirnov
e-mail: Smirnov_A_D@kinef.ru

© Springer International Publishing Switzerland 2016
A. Evgrafov (ed.), *Advances in Mechanical Engineering*,
Lecture Notes in Mechanical Engineering, DOI 10.1007/978-3-319-29579-4_19

183

testing area. As a rule, this area is unit housing. At the same time, other structural elements can actively generate signals. It is particularly typical for units containing internal devices of large sizes, e.g., shell-and-tube heat exchangers, which in this respect are one of the most complex objects of testing. During hydraulic testing of housing, AE signals may come not only from sources (potential defects) in the housing material, but also from sources located in the elements of the tube bundle, which occupies virtually the whole inner space of the unit. Such signals are transmitted via the testing medium (water) to the heat exchanger body and are recorded by the sensors, mounted on it, as location artifact events.

In this case, AE activity areas, non-existent in reality, can be localized on the body, which causes non-productive operations during subsequent inspection of the corresponding body areas by additional NDT methods. They include scaffolding installation or other works in provision of access to the required section, opening and restoration of insulation, dressing of the body for preparation for additional diagnostics, and the diagnostics itself, e.g. by ultrasonic inspection, the result of which will evidently be negative. Such situations reduce the efficiency of AE testing and cause certain authority loss of the method.

This issue can be solved by some methodical techniques of data analysis, allowing for separating of signals from various structural elements. The suggested techniques are implemented under this work on acoustic-emission systems manufactured by Vallen Systeme Company (Germany).

In its turn, solving of the signal separation task makes it possible to use a new approach to solving of the issue of early detection of defects in tubes and other bundle elements. This issue is topical for oil refineries. Thus, for instance, the share of heat exchangers in the structure of vessel equipment of KINEF LLC is 30.7 %, and the share of their failures is 91.8 % (in 2003–2007). At that, the share of through damage of tubes among failure causes is 71.7 %. We suggest using the signals from the unit's inner space, previously assessed as interference during AE testing of the body, as useful information about the presence and location of tube bundle defects.

A method is also suggested for evaluation of AE activity areas with their gradation by degree of potential defect development in order to determine priority in additional testing.

The obtained results can be used in elaboration of AE testing procedures for vessels and units containing internal devices.

Methodical Techniques for Analysis of Location Results of AE Testing

1. The first analysis stage. Separation of location sources by place.

   1.1 First, the task of selection of signal propagation speed for each algorithm (for the shell and for the tube side) is solved. For this the location uncertainty index *LUCY* (*Location Uncertainty*) is applied, which is used in Vallen Systeme systems for analysis of location sources. It is determined for each source and depends on its coordinates, calculated by the location algorithm, and the preset speed of acoustic signal propagation in the medium. It is the value of standard deviation of distances from the i-th sensor to the signal source (S_i), and the same distances calculated according to speed and difference of signal arrival time Dst_i ($Dst_i = \Delta t_i \times V$, Δt_i—*difference of signal arrival time to the 1-st and the i-th sensor, V—signal propagation speed in the medium*):

$$LUCY = \sqrt{\frac{1}{N-1}\sum_{i=1}^{N}(Dst_i - (S_i - S_1))^2} \qquad (1)$$

The value of *LUCY* for the given AE sources is the smaller, the closer the speed, introduced to the location algorithm, is to the real speed of signal propagation from the source to the sensors. In practice, this index can be also interpreted as the radius of a conventional circumference in the area of which the defect should be searched during additional testing.

It is suggested in the procedure of optimal speed selection that the minimum values of *LUCY* should be attained at the best speed. The optimal (the average for the object) speed of signal propagation along the shell can be determined by means of evaluations obtained for the average (for the test) value of *LUCY* for a set of speed values. As a rule, the set of these speeds is generated from the most probable values of propagation of zero modes of Lamb waves (s_0 and a_0) in a shell of known thickness with account of frequency response of the used acoustic emission converters.

Sound speed in the medium inside the heat exchanger can be determined using its value in water. It is virtually constant in the wide range of temperatures and is 1480–1500 m/s. However, the analysis of the result of location of AE sources, located in the tube bundle area, showed that location accuracy increases when calculations use slightly smaller values than the tabular speed value. A probable cause is the extension of the path, along which the signal, passing through the tube bundle, moves from the source to the sensor; the extension is related to deviation of its trajectory from the straight line. Evaluation of the results of such path change allowed suggesting correction factors for the tabular speed value for the standard layouts of tubes' mutual arrangement in the bundle. They were: $k_{45°/90°} = 0.792 - 0.900$ and $k_{30°} = 0.827 - 0.955$. The specific value is set in the calculation for the 3D location algorithm depending on the actual mutual arrangement of AE sensors on the housing and tubes in the bundle.

1.2 The array of recorded data is successively processed by means of planar and 3D location algorithms, assuming the arrangement of AE sources on the housing and inside the unit accordingly.

Then location confidence is evaluated. For this, information is extracted from the data array about the time sequence of signals forming the separate AE events, and on the corresponding distances calculated in the course of operation of location algorithms. The ratios are calculated:

$$R(D_i) = \frac{Dst_{i+1}}{Dst_i}, \quad i = 1\ldots N, N \geq 3 \tag{2}$$

The obtained evaluations allow filtering out the events, for which $(R(D_i)) < 1$ for the given algorithm, as obviously incorrect for the supposed area of their localization. The first separation of AE sources can be performed on this basis. The events, for which this criterion is not met for both algorithms, are removed from subsequent analysis.

1.3 The data, which passed the filter, undergoes at the second stage an evaluation of the degree of correlation between time characteristics (difference of the times of signal arrival to the sensor) and the corresponding distances. For all the signals, included in the events localized by AE, root-mean-square deviations of ratios from the correlation straight line are calculated ($k = 1$):

$$ASD = \sqrt{\frac{1}{N}\sum_{i=1}^{N}(R_i - 1)^2}, \quad i = 1\ldots N, N \geq 3 \tag{3}$$

where $R_i = \frac{R(D_i)}{R(t_i)}$, $R(t_i) = \frac{t_1 + \Delta t_{i+1}}{t_1 + \Delta t_i}$, $t_1 = \frac{Dst_1}{V}$.

The obtained values of ASD parameter are used in several applications. First of all, the region of the most probable localization is determined during comparison of the given characteristic for AE events, for which localization is possible in both regions of the unit. The flow chart of the program implementing this application is shown in Fig. 1.

The program input data is the results of processing of AE testing initial data. Such processing may include different types of filtration and selection by directly measured characteristics and by expert-generated characteristics. Eventually, the results of planar (2D) and three-dimensional (3D) location are generated and used as the program's input data. In the program the information for two location types is combined in order to construct a unified scheme for evaluation of the sources, registered by AE, and their localization in relation to structural elements.

The second application of ASD parameter is the potential possibility of AE source separation by nature of origin if the distribution of AE events for this parameter is characterized by expressed polymodality.

The third application of ASD value, as shown below, is its use as a parameter at the stage of AE activity area evaluation (Fig. 2).

2. The second analysis stage. Procedure for gradation of AE activity sections by sizes of potential defects located within their surface/region.

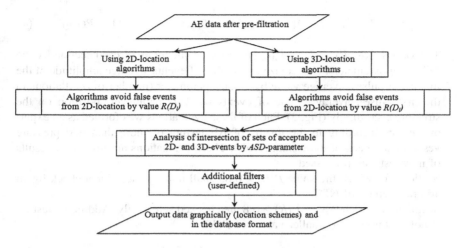

Fig. 1 Flowchart of processing of results 2-D and 3-D location to identify the structural element containing AE sources

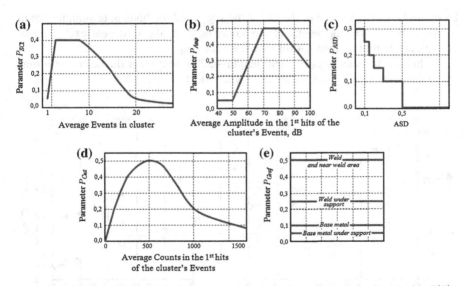

Fig. 2 Set schedules for the calculation of the parameters in the expression for the criterion $P(R)$: **a** P_{SCl}, **b** P_{Amp}, **c** P_{ASD}, **d** P_{Cnt}, **e** P_{Graf}

In practice this evaluation is used for gradation of the revealed AE activity sections by relative expediency (and priority) of additional testing. For this the $P(R)$ criterion is used, expressed as:

$$P(R) = 1 - (1 - P_{SCl})(1 - P_{Amp})(1 - P_{Cnt})(1 - P_{ASD})(1 - PGraf) \qquad (4)$$

The values P_{SCl}, P_{Amp}, P_{Cnt}, P_{ASD}, P_{Graf} are evaluations of each section by the following parameters: cluster size by number of events, average amplitude of the first event pulses, average value by Counts parameter (number of oscillations in the first pulse), average value of events by ASD parameter, position on the structure accordingly (Fig. 3). Each of these evaluations was formed as an expert evaluation according to the results of AE testing of more than 600 pressure vessels and check of the testing results. These evaluations are revised as results of new tests are processed.

Sections for which the value $P(R) \geq 0.85$ shall undergo additional check by an arbitral method of NDT.

Sections for which $0.85 \geq P(R) \geq 0.8$ are checked selectively. Additional testing is not performed for smaller values of $P(R)$.

Examples of Use of the Suggested Methical Techniques

Example 1. As an example of the use of the suggested methical techniques, let us consider the AE testing results of the lower body of the feed heat exchanger of preliminary hydrotreatment of oil refinery reforming.

Technical characteristics of shell: length—5280 mm; inner diameter—600 mm, wall thickness—16/25 mm, material—carbon steel 16GS.

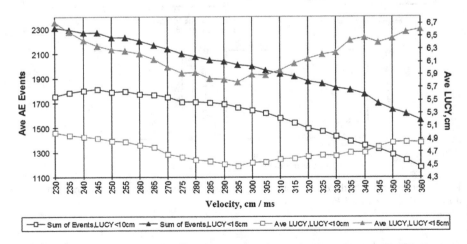

Fig. 3 Relation of the average value $LUCY$ and the number of localized events from the velocity of acoustic signal propagation in the shell

The result of evaluation for the optimal value of signal propagation speed along the unit housing is shown in Fig. 4. The graphs of dependence of the average *LUCY* value on the speed, used in calculations, are minimum at $V = 295$ cm/ms for different levels of data filtration by *LUCY* value.

Figure 5 on the developed view of a body's cylindrical surfaces (outer view) shows the result of the use of ratios of kind (1) for separation of AE sources into

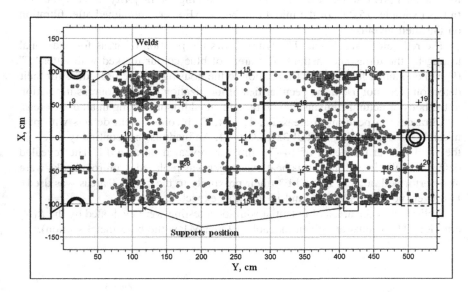

Fig. 4 Isolation false AE sources (indicated by *red* symbols) among the formal results of the planar location at shell's reamer of the cylindrical surface of the heat exchanger (outside view)

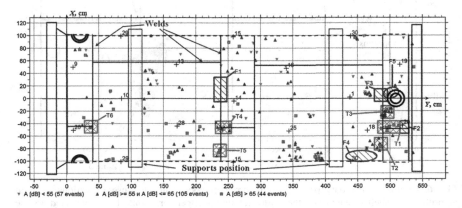

Fig. 5 Results of planar location after excluding a core set of false events from common data. Position for 6 sites (with the letter "T" in the designation) for arbitration inspection is shown

true and false events. About 33.4 % of false sources on the housing were filtered out at this stage (453 of 1356). The diagram shows the precise position of welded joints, supports and connection pipes. This allows using it as a full-scale graphic filter during analysis of AE activity source position.

Then the data was evaluated according to the *ASD* parameter. Figure 6 shows the results of data processing after omission of the main array of artifact events. The used approach does not obviously reveal all the false events, but it is rather efficient for practical AE testing, because it allows separating the majority of such sources. In this example, 15.2 % of the initial number of AE events remained after filtration on the shell layout.

The red colour and letter "T" (True) shows the position of areas for additional testing by the ultrasonic method. The areas of blue colour marked with letter "F" (False) contained false events (in relation to the housing). If the procedure for their detection were not used, the peculiarities of localization of such sources in relation to the body structural elements would require an additional testing in all these areas.

Ultrasonic inspection on areas T2, T3, T4, T5 recorded weld defects with permissible sizes according to RDI 38.18.016-94. An inadmissible continuity defect in the weld root at the depth of 14–21 mm was revealed on area T1. The revealed defect was eliminated, later this heat exchanger was replaced. Control checks were also performed in areas F2 and F3 (their proximity to the group T areas was used). However, no defects were found in them, as expected.

Preliminarily the areas of potential additional testing were evaluated by the $P(R)$ criterion. The results are given in Table 1 (area M1 is the base metal section).

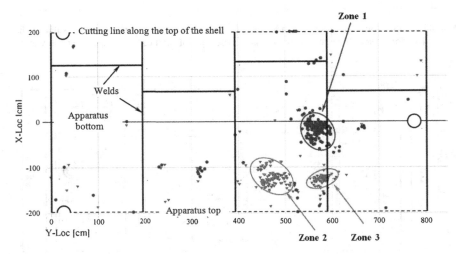

Fig. 6 Results of the data processing by program coordinates the use of various location algorithms at shell's reamer of the hydro-refining refrigerator. Zone 1 is connected to the registration AE sources located in the shell; generation of Zone 2 and Zone 3 is connected with the registration of AE activity in the elements of the tube bundle

The data of Table 1 shows that the results of additional testing are well in keeping with the evaluation of AE activity areas by the $P(R)$ criterion.

Example 2. AE testing of the housing of gasoline hydrotreater feed cooler. Technical characteristics. Shell: length—8040 mm; inner diameter—1200 mm, wall thickness—30 mm, housing material—carbon steel 16GS. Tube bundle: number of tubes—408 pcs, tube inner diameter—25 mm, thickness—2 mm, material—carbon steel St20.

This example illustrates the applications of the program for identification of AE signals generated by various structural elements. The results of its work made it possible to separate the sources located on the housing and related to internals. Area 1 on the body developed view (Fig. 6) is classified as the area of true AE events in

Table 1 Results of additional testing

Area denotation	Sum AE events	Average			P_{Graf}	$P(R)$
		Amplitude	Counts	ASD		
T1	6	78.2	246	0.13	0.5	0.94
T2	6	63.2	142	0.08	0.5	0.89
T3	12	61.0	107	0.06	0.5	0.86
T4	9	57.4	69	0.07	0.5	0.85
T5	4	66.2	134	0.10	0.5	0.90
T6	2	67.7	158	0.06	0.5	0.89
M1	17	55.4	61	0.07	0.1	0.38

Fig. 7 Defects found in the area of AE activity corresponding to Zone 1: pitting corrosion inner surface of the heat exchanger shell

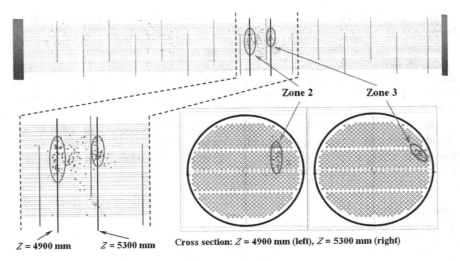

Fig. 8 Position of zones AE activity in the area of the tube bundle refrigerator associated with corrosion damage of two groups of closely spaced tubes

Fig. 9 Zone tube bundle corrosion damage (Zone 3)

relation to the body. It corresponds to the location of the area of significant corrosion damage and pit corrosion of the housing on the inside (photo in Fig. 7).

Areas 2 and 3 in relation to the body are areas of location artifact clusterization, but they are "projections" of the real AE events, generated in the unit internal device, onto the body. Figure 8 shows the localization of these events in the bundle at 3D location.

During AE testing, there were here areas of corrosion damage of several adjacent tubes, but without formation of a through defect. Later, due to development of corrosion processes under the impact of operational factors, perforation of several tubes occurred, the heat exchanger failed and the bundle had to be urgently replaced. Figure 9 shows the state of tubes at the time of bundle rejection on the section corresponding to Area 3 in Fig. 8.

Conclusions

The suggested methodical techniques for data analysis, used for AE source separation by their location in process units containing internals, and for gradation of AE activity areas provide the following:

1. Increase confidence of AE testing of heat exchange equipment bodies (reliability of nondestructive testing is an index related to the probability of error-free decisions on presence or absence of defects), and, consequently,
2. Reduce the sections of additional testing according to AE results and, accordingly, time and material expenses on the preparation and execution of such works.
3. Consider the prospects of AE method use for localization and early diagnostics of tube bundle elements in heat-exchange equipment according to the results of simultaneous integrated AE testing of units.

The suggested approach to analysis of heat exchange AE testing data can be applied to other types of process units, also comprising internals (columns with trays, electric dehydrators etc.). It can be used for creation of AE testing procedures for such equipment.

References

1. Greshnikov VA, Drobot JB (1976) Acoustic emission M. Standards Publishing House, 272 p
2. Rules of the method of acoustic emission application during inspection of boilers, vessels, apparatus and process piping. PB 03-593-03 M (2003)
3. Ivanov VI, Belov IE (2005) The method of acoustic emission. In: Klyuev VV (ed) Reference, vol 7, M: machine-building, 825 p
4. Nefedyev EJ (1986) Connection sizes of microcracks with the acoustic emission parameters and structure of deformed steel rotor. Defectoscopy, no 3, pp 41–44
5. Nefedyev EJ (1987) The control of the fatigue cracks growth in cast steel method of acoustic emission. Problems of strength, no 1, pp 41–44
6. Kovalev DN, Nefedyev EJ, Tkachev VG (2012) Acoustic emission control testing of steel corrugated pipes of circular and static loading. Modern engineering. In: Radkevich MM,

Evgrafova AN (ed) Science and education: materials of the 2nd international nauch—practical use conference. SPb, Publishing house Polytechnic University, pp 382–390

7. Nefedyev EJ (2013) The use of acoustic emission method with spectral analysis of signals to determine the parameters of a leak in the pipeline ITER. Modern engineering. In: Radkevich MM, Evgrafova AN (ed) Science and education: materials of the 3rd international nauch—practical use conference. SPb, Publishing house Polytechnic University, pp 347–355

8. Gomera VP, Sokolov VL, Fedorov VP (2008) Implementation of acoustic emission method to the conventional NDT structure in oil refinery. In: Journal of acoustic emission, Acoustic Emission Group, Ehcino, CA, USA, vol 26, pp 279–289

9. Kabanov VS, Sokolov VL, Gomera VP, Fedorov VP (2011) The Experience of application of acoustic emission technique for the inspection of pressure vessels. Chemical engineering. № 4, pp 12–15

10. Catty J (2009) Acoustic emission testing–defining a new standard of acoustic emission testing for pressure vessels. Part 1: quantitative and comparative performance analysis of zonal location and triangulation methods. J Acoust Emission 27:299–313

Author Index

A
Aksenov, Leonid B., 175

B
Bahrami, Mohammad Reza, 67
Belikov, Victor, 149
Borina, Anastasia, 127

C
Chekanin, Alexander V., 87
Chekanin, Vladislav A., 87

E
Eliseev, Kirill V., 57
Eliseev, Vladimir V., 93
Evgrafov, Alexander N., 31

F
Filippenko, George V., 115

G
Galiullin, Ilnar A., 17
Gavrilov, Andrey, 135
Gomera, Victor P., 183

I
Ivanov, Evgeniy K., 159

K
Karazin, Vladimir I., 75
Khlebosolov, Igor O., 75
Kovalev, Mikhail D., 1
Kozlikin, Denis P., 75
Krylov, Nikolay A., 159
Kunkin, Sergey N., 175

M
Mingazov, Marat R., 9
Moskalets, Artem A., 93

N
Nefedyev, Evgeny J., 183
Nesmiynov, Ivan, 149

O
Oborin, Evgenii A., 93

P
Petrov, Gennady N., 31

S
Shamanov, Ilya, 135
Skakunov, Vladimir, 135, 149
Skotnikova, Margarita A., 159
Smirnov, Anatoly D., 183
Sukhanov, Alexander A., 39, 75

T
Tereshin, Valerii, 127
Tsvetkova, Galina V., 159

Y
Yarullin, Munir G., 9, 17

Z
Zhoga, Victor, 135, 149
Zhukov, Vladimir A., 169
Zlatanov, Vassil, 105

© Springer International Publishing Switzerland 2016
A. Evgrafov (ed.), *Advances in Mechanical Engineering*,
Lecture Notes in Mechanical Engineering, DOI 10.1007/978-3-319-29579-4

Printed in the United States
By Bookmasters